Arndt • Haenel, π – Unleashed

Springer
*Berlin
Heidelberg
New York
Barcelona
Hong Kong
London
Milan
Paris
Singapore
Tokyo*

Jörg Arndt   Christoph Haenel

Pi – Unleashed

Translated from the German
by Catriona and David Lischka

With CD-ROM

 Springer

Jörg Arndt
Hühlweg 37
95448 Bayreuth, Germany

Christoph Haenel
Waldfriedenweg 24
82223 Eichenau, Germany

Title of the original German edition:
Pi. Algorithmen, Computer, Arithmetik
© Springer-Verlag Berlin Heidelberg 1998, 2000

Library of Congress Cataloging-in-Publication Data applied for
Die Deutsche Bibliothek – CIP-Einheitsaufnahme
Arndt, Jörg: Pi-unleashed: with CD-ROM / Jörg Arndt; Christoph Haenel.
Transl. from the German by Catriona and David Lischka. – Berlin; Heidelberg;
New York; Barcelona; Hong Kong; London; Milan; Paris; Singapore; Tokyo:
Springer, 2001
Dt. Ausg. u.d.T.: [Pi]
ISBN 3-540-66572-2

ISBN 3-540-66572-2 Springer-Verlag Berlin Heidelberg New York

This work consists of a printed book and a CD-ROM packaged with the book, and is subject to copyright. All rights are reserved, whether the whole or part of the material is concerned, specifically the rights of translation, reprinting, reuse of illustrations, recitation, broadcasting, reproduction on microfilm or in any other way, and storage in data banks. Duplication of this publication or parts thereof is permitted only under the provisions of the German copyright law of September 9, 1965, in its current version, and permission for use must always be obtained from Springer-Verlag. Violations are liable for prosecution under the German Copyright Law.

Springer-Verlag or the authors make no warranty for representation, either express or implied with respect to this CD-ROM or book, including their quality, merchantibility, or fitness for a particular purpose. In no event will Springer or the authors be liable for direct, indirect, special, incidental, or consequential damages arising out of the use or inability to use the CD-ROM or book, even if Springer-Verlag or the authors have been advised of the possibility of such damages.

Springer-Verlag Berlin Heidelberg New York
a member of BertelsmannSpringer Science+Business Media GmbH
© Springer-Verlag Berlin Heidelberg 2001
Printed in Germany

The use of general descriptive names, trademarks, etc. in this publication does not imply, even in the absence of a specific statement, that such names are exempt from the relevant protective laws and regulations and therefore free for general use.

Cover design: Künkel + Lopka Werbeagentur, Heidelberg
Typesetting: Camera-ready by the authors
Printed on acid-free paper – SPIN: 10731352   33/3142 GF 5 4 3 2 1 0

Nil desparare
(Do not despair)

Motto from Gauss's Mathematical Diary (1796–1814),
in which the author demonstrates a connection between
the computation of π and the arithmetic-geometric mean.

Foreword to the second edition

After the first edition of our book on the fascinating number π was so well received, we are pleased to present the second edition, which has been completely revised. All the various subject areas have been expanded, improved and updated. We are grateful to those of our readers whose suggestions we have adopted.

The update includes a report on the latest world record by the Japanese Professor Kanada, who has now calculated π to 206.1 billion decimal places, and also the record achieved by Colin Percival, who in 1999 while still a teenager calculated the 40 trillionth binary digit of π with the assistance of a team of Internet collaborators. In addition there is also a new section on *FFT multiplication* (multiplication using fast Fourier transforms), in which this technique that today is indispensable for modern π calculations, is explained in detail with appropriate examples of code.

The historical section has been completely reworked and considerably enlarged.

We have also expanded the (platform-independent) CD-ROM to include, in particular, 400 million decimal places of π and several freeware high precision libraries. Last but not least, the URLs for many π sites on the Internet have been brought up-to-date.

We would particularly like to thank Professor F.L. Bauer of the Technical University of Munich for his invaluable ideas and suggestions for improvements on every topic in the book. We feel honoured to have enjoyed Professor Bauer's encouragement and advice.

We look forward eagerly to the reactions and verdict of our readers. We now have a special e-mail address to which you can send your comments. Please write to **pibook@jjj.de**.

Have fun!

Jörg Arndt and Christoph Haenel

Foreword to the first edition

The number π, about which most people know only that it is expressed as the ratio of the circumference of a circle to its diameter and begins with 3.14..., was calculated in July 1997 to 51.5 *billion* decimal places. In this book you will discover why anyone should be interested in achieving world records with such bizarre calculations, which are long enough to fill up thousands of phone directories, and you will also find many other interesting things about this number. The methods which are used for these high-precision calculations reflect developments in mathematics whose significance goes far beyond simply "working out π". These new and effective algorithms surprisingly have the pleasant characteristic that they are easy to understand.

In this book we describe in detail the quest for world records in π and on the enclosed CD-ROM we provide programs which will enable you yourselves to calculate several million digits of π on your computer. The CD-ROM also contains source code versions of all the programs and examples, together with more detailed text on topics which had to be left out of the book due to space constraints. To help you get started with your own research on the Internet, we have also provided a collection of URLs. For the statisticians amongst you, we have included several million digits of π on the CD-ROM.

Finally we narrate the main personalities and milestones in the history of π, amongst which you will find not only facts of scientific interest but also some examples of both the curious and the bizarre.

The following persons provided extremely useful comments and suggestions as we were writing the book: Dieter Beule, Lothar Krüger, Heinz Pöhlmann, Georg Sessler, Mikko Tommila, Ute Zwerschke plus a particularly co-operative reader who wishes to remain anonymous. We would like to express our heartfelt thanks to all of them.

We wish you a pleasant journey into the fascinating world of the number π .

August 2000
Jörg Arndt and Christoph Haenel

Contents

1. **The State of Pi Art** 1
2. **How Random is π?** 21
   - 2.1 Probabilities 21
   - 2.2 Is π normal? 21
   - 2.3 So is π not normal? 24
   - 2.4 The 163 phenomenon 25
   - 2.5 Other statistical results 28
   - 2.6 The Intuitionists and π 30
   - 2.7 Representation of continued fractions 32
3. **Shortcuts to π** 35
   - 3.1 Obscurer approaches to π 35
   - 3.2 Small is beautiful 37
   - 3.3 Squeezing π through a sieve 38
   - 3.4 π and chance (Monte Carlo methods) 39
   - 3.5 Memorabilia 44
   - 3.6 Bit for bit 47
   - 3.7 Refinements 49
   - 3.8 The π room in Paris 50
4. **Approximations for π and Continued Fractions** 51
   - 4.1 Rational approximations 51
   - 4.2 Other approximations 55
   - 4.3 Youthful approximations 63
   - 4.4 On continued fractions 64
5. **Arcus Tangens** 69
   - 5.1 John Machin's arctan formula 69
   - 5.2 Other arctan formulae 72

6. Spigot Algorithms ... 77
6.1 The spigot algorithm in detail ... 78
6.2 Sequence of operations ... 80
6.3 A faster variant ... 82
6.4 Spigot algorithm for e ... 84

7. Gauss and π ... 87
7.1 The π AGM formula ... 87
7.2 The Gauss AGM algorithm ... 90
7.3 Schönhage variant ... 92
7.4 History of a formula ... 94

8. Ramanujan and π ... 103
8.1 Ramanujan's series ... 103
8.2 Ramanujan's unusual biography ... 105
8.3 Impulses ... 110

9. The Borweins and π ... 113

10. The BBP Algorithm ... 117
10.1 Binary modulo exponentiation ... 120
10.2 A C program on the BBP series ... 123
10.3 Refinements ... 126

11. Arithmetic ... 131
11.1 Multiplication ... 131
11.2 Karatsuba multiplication ... 132
11.3 FFT multiplication ... 135
11.4 Division ... 145
11.5 Square root ... 146
11.6 nth root ... 149
11.7 Series calculation ... 150

12. Miscellaneous ... 153
12.1 A π quiz ... 153
12.2 Let numbers speak ... 154
12.3 A proof that $\pi = 2$ ... 155
12.4 The big change ... 155
12.5 Almost but not quite ... 156
12.6 Why always more? ... 158

12.7 π and hyperspheres 158
12.8 Viète × Wallis = Osler 160
12.9 Squaring the circle with holes 162
12.10 An (in)finite funnel 164

13. The History of π 165
13.1 Antiquity 167
13.2 Polygons 170
13.3 Infinite expressions 185
13.4 High-performance algorithms 198
13.5 The hunt for single π digits 203
Table: History of π in the pre-computer era 205
Table: History of π in the computer era 206
Table: History of digit extraction records 207

14. Historical Notes 209
14.1 The earliest squaring the circle in history? 209
14.2 A π law 211
14.3 The Bieberbach story 213

15. The Future: π Calculations on the Internet ... 215
15.1 The binsplit algorithm 215
15.2 The π project on the Internet 219

16. π Formula Collection 223

17. Tables 239
17.1 Selected constants to 100 places (base 10) 239
17.2 Digits 0 to 2,500 of π (base 10) 240
17.3 Digits 2,501 to 5,000 of π (base 10) 241
17.4 Digits 0 to 2,500 of π (base 16) 242
17.5 Digits 2,501 to 5,000 of π (base 16) 243
17.6 Continued fraction elements 0 to 1,000 of π .. 244
17.7 Continued fraction elements 1,001 to 2,000 of π .. 245

A. Documentation for the hfloat Library 247
A.1 What hfloat is (good for) 247
A.2 Compiling the library 248
A.3 Functions of the hfloat library 248
A.4 Using hfloats in your own code 250

A.5 Computations with extreme precision 250
A.6 Precision and radix 251
A.7 Compiling & running the π-example-code 253
A.8 Structure of hfloat 253
A.9 Organisation of the files 254
A.10 Distribution policy & no warranty 255

Bibliography 257

Index 265

1. The State of Pi Art

In October 1999 Professor Yasumasa Kanada of the University of Tokyo announced on the Internet[1] that he had set a new world record for the calculation of π; namely, he had calculated it to 206,158,430,000 decimal places. He had performed two independent calculations using two different algorithms, each of which had produced identical results except for the last 45 digits. The programs used to perform the calculations were written by Kanada and his assistant, Daisuke Takahasi. They had taken 83.5 hours of computer time to execute.

With this latest breakthrough, humankind now possesses 206.1 billion digits of π beginning with 3.1415. That is indeed a lot of numbers. Just to have them race past on the screen would take several weeks. If one were to print out 100 million digits from this π (on around 10,000 sheets of paper), there would still be 206 billion digits left. They would occupy 30,000 books, each 1000 pages long, although equally they would also fit on just a few hard disks each weighing less than a pound.

Yasumasa Kanada's statement came only five months after the same man announced he had set a new world record of 68.7 billion decimal places. That announcement in turn came only 18 months after reaching 51.5 billion digits. Since 1981, when the record stood at 2 million digits, it has been beaten 26 times, almost doubling every year.

For some years Kanada has been the only π mathematician intent on achieving new world records. Up until 1986, when a comparatively modest 30 million digits was still a world record, a number of competitors were still in the arena. Ten years later this had been whittled down to two, Kanada in Japan and the Chudnovsky brothers in the USA, who took it in turns to outdo each other until in 1996 the 8 billionth

---

[1] ftp://pi.super-computing.org/README.our_latest_record

decimal place was reached. Since 1997 and from 51 billion digits, only Kanada and his team are still in the running.

The computers used to attain these world records have generally been supercomputers which were specially developed for number-crunching and cost millions of dollars. For example, Kanada achieved his latest record on a Hitachi SR8000 with 128 processors. Nevertheless, if one has the skill it is possible to make smaller computers perform π calculations which are still competitive. In the history of the world records there have been several examples of this. Thus, for example, a few years ago the above-mentioned Chudnovsky brothers calculated their 8 billion decimal digits of π in New York on a computer which they had built themselves from parts purchased in department stores [92].

One might expect Kanada to record his π calculations in compressed form[2] on around 150 CDs and to offer them for sale to interested parties. However, the professor seriously views his π as "non-commercial digits" and will only release them after checking carefully that they are to be used for scientific purposes. He will only allow the first few billion digits to be downloaded over the Internet from his server in Tokyo. This itself is not a trivial task since, with a normal telephone line, downloading the data would take approximately one day per billion digits (with the associated expense). If the explosion in the capability of computers continues, it will soon be quicker and simpler to calculate one's own π to a suitable length than to download it from elsewhere. When one is dealing with a few hundred million digits this is in fact already the case today. And even a sophisticated π program will always be shorter than the number of digits to which it can calculate.

What is the point of having such a long representation of π? One obvious task is to search through the digits looking for rare or unusual patterns and possibly for lengthy sequences of the same digits. For the first few thousand digits, which are shown on page 240ff. of this book, it is probably still feasible to do this without a computer. Above that limit, however, one will probably need a search program, for example our `piseek` program, which is included on the enclosed CD-ROM along with the first 400 million digits of π and should provide a few hours' pleasure.

---

[2] Because the decimal digits of π are (probably) normally distributed, at best N digits would compress down to $1/(8\log_{10} 2) N$, around $0.4 N$ bytes.

As one examines π, one is struck by a striking feature which occurs as early as at position[3] 768, where 9 occurs six times in succession. This peculiarity, the probability of which is only 0.08%, even has a name. It is known as "Feynman point", after the American Nobel prize winner Richard Feynman (1918–1988). He once claimed that if he had to recite the digits of π, he would name them accurately up to this point and only then say *and so on*. His words have an elegant double meaning, since after these six 9s it is of course *not* at all a case of "and so on", but something quite different follows, in this case an 8. The next block of six identical numbers occurs only much later, but still 'too early', at position 193,034. Remarkably, it consists of a second series of 9s.

On the other hand, the first zero in π does not occur until position 32, i.e. much later than one would expect. These properties of the number π have facilitated the task of "π poets" seeking to compose aids to remembering π in which every word has the same number of letters as the related π digit. A zero would terminate such a sequence *by definition*. One of these mnemonic verses, which produces a lot of coffee and 11 pieces of π, runs as follows:

May I have a large container of coffee? Cream and sugar?
3   1   4   1   5   9   2   6   5   3   5

Where does the sequence 0123456789 first appear in π? It was only in 1997, at the time of the last but one world record, that this obvious question was finally answered. In fact that sequence does not occur until position 17,387,594,880, i.e. long past the 6.4 billionth digit, which was previously the furthest that was known. But, if this special sequence does not occur until that late, another special sequence, consisting of 10 digits all different from each other, occurs relatively early, commencing in position 60. Even in the 17th century, mathematicians had calculated π this far, as the history of π world records shows on page 205.

Number fanatics might perhaps search in π for "self-referential positions", in which the numeric sequence is the same as the position number. The first such position occurs right at the beginning, namely in position 1. Other examples are to be found at positions 16,470 and 44,899, but after that there are no further instances until posi-

---

[3] It is customary to count digit positions from 0, beginning with the 3 in front of the decimal point. In this way 3 1 4 1 5 correspond to positions 0 to 4.

tion 79,873, 884. Whether there are any self-referential decimal places after that we do not know.

It may be more communicative to search through π for particular telephone numbers or other numbers such as one's own birth date, as people have always been intrigued by the question, *Where do I occur in π?* (For example, the author's birth dates are at positions 5,407,560 and 14,666,671). If I liked, I could print the position on my business cards, communicating vital information about what kind of person I am. On the Internet there is a game like this, although currently it is only based on 50 million digits[4].

Somewhere in π, *everything* that is finite is surely to be found, including, when coded in numeric form, every text in the world, the shortest, the longest, the cleverest. Naturally that includes this text, and also of course the Bible, in every language, and every piece of music. (Moreover, π also contains π itself, albeit only once, whereas neither $\pi + 1$ nor the root of 2 will be found within π.)

Human DNA contains about 3.6 billion digits of information, so my neighbour's DNA could already be contained in the known digits of π.

Perhaps, but highly improbable. For example, the number which Y. Kanada has calculated does not even contain all the 20-digit numbers that are possible, which is 10^{20}, i.e. far more than the approximately 10^{11} digits which occur in Kanada's number. Still less does it contain all the 21-digit numbers, let alone all the numbers of DNA or Bible length. This makes it unlikely that the Bible, with its estimated 10^7 digits, occurs in the first "few" digits of π; to find the Bible, we would need to possess around 10^{10^7} digits, whereas in fact we have little more than 10^{10^1}.

Even worse, it simply is not possible for theoretical reasons to calculate π to so many digits, for the universe consists only of 10^{79} elementary particles, and even if one were to turn the entire universe into a supercomputer, it would only be able to store about 10^{76} decimal places, which is far fewer than the 10^{10^7} Bible digits mentioned above. However, it is conceivable that individual positions on this side of that limit can be calculated.

Anyone who wants to make a name for himself can examine the major issue of whether π is *normal*, or perhaps more accurately, whether π is *not normal*. If π were not normal, this would mean that particular

---

[4] http://www.aros.net/~angio/pi_stuff/piquery

numbers or blocks of numbers of the same length occur more often or less often than others, so that, for example, in the 50th billion positions significantly more sixes occur than in the 60th billion.

Anyone who succeeded in making such a pronouncement would certainly make the headlines. For up to now no one has found any sections within π which are non-normal. Attempts to represent π in hexadecimals, under which π begins with 3.243F6 A8885, rather than in the customary decimal representation, have likewise been unsuccessful. Even the so-called simple continued fraction of π, i.e. the representation of π using an infinite series of fractions in which all the numerators are equal to 1,

$$\pi = 3 + \cfrac{1}{7 + \cfrac{1}{15 + \cfrac{1}{292 + \cfrac{1}{\ddots}}}}$$

under which the fine structure of numbers becomes evident, has not revealed any regularities, although the simple continued fractions of some other transcendental numbers, such as the number e, indeed do exhibit recognisable patterns.

Mathematicians conclude from these observations and from other indications that π is a normally distributed number. But to date they have not been able to prove this, so it is entirely possible that "somewhere" within π a big sensation is waiting to be uncovered. This question is considered further in the next chapter.

It is thus unclear whether π is normal. On the other hand, it is definitely known to be irrational, as was demonstrated for the first time in 1766 by the Alsatian mathematician, Johann Heinrich Lambert (1728–1777). An irrational number is one which cannot be represented as a ratio of two integers. For example, the fraction $\frac{355}{113} = 3.141\,592\ldots$ is a very good approximation as it specifies π to 6 correct decimal places; nevertheless, it is not *equal to* π, and no other fraction having an integer numerator and denominator no matter how long can produce the exact value of π. The approximation $\frac{355}{113}$ was discovered already in the 5th century AD by Tsu Chhung-Chih in China, and for almost 800 years was the best π approximation available. This and other approximations are considered in Chapter 4.

It is also known that π is transcendental. A famous, but extremely complicated proof of this was provided by the Munich mathematician Ferdinand Lindemann (1852–1939) in 1882, i.e. more than 100 years after Lambert's conclusions as to the irrationality of π. Lindemann's proof states that π cannot be the root of a polynomial with integral coefficients, however many terms it may have. It is true, for example, that $9\pi^4 - 240\pi^2 + 1492$ is *approximately* 0 (or, more precisely, $-0.02323\ldots$), but it is impossible to find such an expression which gives 0 *exactly*.

It is astonishing but nevertheless true that today, several hundred years after the transcendence of π was proven, we do not know much more about π.

Admittedly, the constants π^2, e^π and $\pi + \log 2 + \sqrt{2}\log 3$ have now also been proven to be transcendental. But we still do not know whether similar quantities such as $e + \pi$, $e \cdot \pi$, π/e, $\log \pi$ or π^e are irrational or not. Many other questions regarding π remain unanswered. Some questions, such as the issue of whether π is normal or not, are definitely unresolved, because no one has an inkling as to what method one might employ in order to resolve the matter one way or the other. And when we consider the 206.1 billion digits that are known, all we know about the decimal sequence of π is trivial information such as the frequency and distribution of (short) blocks of numbers [11, p. 203].

The number π is thus one of the oldest subjects of research by mankind and possibly the one topic within mathematics which has been researched for the longest. Humans have concerned themselves with π for several thousand years. For example, already in 2000 BC the Babylonians and Egyptians had discovered approximations for π which were less than 0.02 away from its true value. Over the next, increasingly enlightened 4000 years, researchers tirelessly never gave up trying to unravel the secrets of π. The fascinating history of this research is covered in more detail in Chapter 13, page 165ff.

When one considers all the work that has been done on π, the meagre amount that is known about it is somewhat surprising. But then when one considers that scarcely no other field is quite like number theory in throwing up very simple questions that are so difficult to answer, perhaps it is not so surprising after all. One example here is the so-called Last Theorem of Pierre de Fermat (1601–1665) of 1637, which states that there are no integer solutions for the equation

$$x^n + y^n = z^n$$

if n is greater than 2. This old and easy-to-state problem remained unresolved for centuries although countless numbers of (more or less) educated people endeavoured to resolve it; it was only recently (October 1994) and with a complicated 100-page proof that Fermat's Last Theorem was finally proven by Andrew Wiles. On the other hand, problems of number theory which are also easy to formulate always turn out to defeat all but the very best minds, for example the famous question of whether the number of prime pairs, i.e. sequences of two prime numbers separated by 2 such as the sequence 3 and 5 or 10,007 and 10,009, is infinite. Simple questions - yet no one has solved them yet.

With his proof of the transcendence of π, Lindemann also settled another problem which has preoccupied the brains of the best mathematicians and philosophers since the ancient Greeks and which was already so well-known in 414 BC that Aristophanes made use of it in his comedy "The Birds".

This is the problem of the *squaring the circle*, i.e. the challenge of finding a square which has exactly the same area as a given circle using a geometrical construction whose execution requires nothing but an unmarked straight edge and compass and must be possible in a finite time. With r as the radius of the circle, one needs to construct side x of a square whose area x^2 is equal to the area of the circle $r^2\pi$. The sides of the square must therefore equal $x = r\sqrt{\pi}$.

But only certain line lengths can, under the specified conditions, be determined through geometric constructions, namely ones which are integers and are produced through normal, rational operations such as addition or division and/or through the formation of square roots [32, p. 347]. $\frac{8}{3}\sqrt{17}$, for example, can be constructed in this way, whereas $3\sqrt[3]{3}$ cannot. The expression $r\sqrt{\pi}$ does contain a multiplication and a square root, and hence meets the above requirement for a solution to the problem. But to allow the circle to be squared, it would have to be possible to obtain the radicand π through the specified operations, and it is precisely this that does not occur according to Lindemann's proof. Hence, it is not possible to square the circle with a geometric construction using only compass and ruler. (On the other hand it is possible if the constraining conditions are varied. For example, in hyperbolic geometry there may well be a circle, i.e. one whose area is equal to π, which can be transformed into a square with the same area with the aid of (unmarked) compass and straight edge; but in this - non-Euclidean - geometry a straight line cannot be drawn with a ruler.)

After Lindemann had resolved the problem of the Squaring the circle once and for all in 1882 (even if his achievement was only to demonstrate that the problem cannot be resolved), one would have expected all further attempts at a solution to cease immediately. It is all the more amusing to find that despite Lindemann's proof a lot of people continued to try to solve the problem, and they are still at it even today. Dudley writes about such "mathematical cranks" in a monograph [50], in which he mentions amongst others the ship builder O.Z. (Otto Zimmermann) from Hamburg who published a book entitled "π is rational" in 1983. In his book Dudley also classifies those who have attempted to square the circle. According to this, O.Z. is a member of that relatively rare species who keep revising their opinion over time: before 1975 his construction produced $\pi = 3.14159\,26535\,576$, then up to January 1976 it was $3.14159\,26535\,98$, while from then on π amounted to exactly 3.1428.

π is the ratio of the circumference c of a circle to its diameter d; $\pi = c : d$. This is the classic geometrical definition of the number. The popular rule that π is the circumference of a circle with diameter 1 is based on this.

A second geometrical definition states that π is the ratio of the area of a circle A to the square of the radius r of the circle i.e. it is determined by the relationship $\pi = A : r^2$. This results in the obvious statement that the ratio of the area of the circle to the area of the surrounding square is given by $\pi : 4$.

The oldest Indo-Germanic cultures were already aware, thousands of years BC, that a fixed ratio exists between the circumference and the diameter of all circles, large or small. The second fixed ratio in the circle, namely the relationship between the area of the circle and the square of the radius, was also known unimaginably long ago, even though this is less obvious than the first. On the other hand, it was considerably later, perhaps not until ancient Greece, that it was appreciated that both these relationships involve the same ratio, π. This fact is not readily understood. The picture below is perhaps the simplest way of examining the concept:

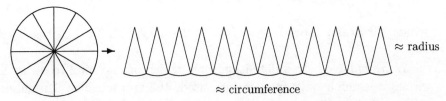

In the picture on the left, a circle has been divided into a number of equal-sized "slices of cake". If one imagines that the slices were taken out and arranged side by side in a rectangle, they would appear as shown in the picture on the right. The greater the number of slices, the more the outline of the rectangle approximates the length of the circumference of the circle, while the height of the rectangle approaches the length of the radius of the circle. The rectangle in the picture is only half occupied by the cake slices, and is therefore twice as big as the sum of their areas, which is equal to the area of the circle. Thus, if π appears in the formula for the area of the circle, then it also figures in the calculation of the circumference.

In addition to the geometric definition of π, there are many other definitions, including ones drawn from quite different contexts. For example, π appears in interesting probability problems.

What is the probability that a coin tossed an even number of times will land an equal number of times on its head and on its tail? From one's schooldays one may recognise the formal treatment of such exercises: if the coin is tossed $2n$ times and there are two possible outcomes per toss, then altogether there are 2^{2n} possible outcomes, of which $\binom{2n}{n}$ will be the desired outcome. The probability of this is therefore $p = \binom{2n}{n}/2^{2n}$, or

$$p = \frac{1 \times 3 \times 5 \times \ldots \times (2n-1)}{2 \times 4 \times 6 \times \ldots \times 2n} \qquad (1.1)$$

This fraction is difficult to calculate for large values of n. We have John Wallis (1616–1703) to thank for calculating it before infinitesimal analysis was invented. Following an interesting train of thought, he discovered that the infinite product comes to

$$\frac{3 \times 3 \times 5 \times 5 \times 7 \times 7 \times \ldots}{2 \times 4 \times 4 \times 6 \times 6 \times 8 \times \ldots} = \frac{4}{\pi} \qquad (1.2)$$

It is not difficult to see [123] that this *Wallis product* occurs in the formula for the calculation of p shown above, and in such a manner that

$$p \approx 1/\sqrt{\pi \cdot n} \qquad (1.3)$$

which, for example produces a 10% probability with $2n = 62$ tosses of the coin.

We thus have here a definition for π which has nothing to do with circles; although only integers are used, the transcendental number still appears in the formula.

π figures in many other probability statements as well. In particular it occurs in the well-known Gauss error distribution curve, which, for instance, is used in graduated life tables for life assurance. Some bright spark [92] remarked à propos of this that π even occurs in death.

There are definitions of π which are even more abstract, for example, analytical definitions using integrals. Two particularly visually attractive examples were produced by the Swiss mathematician Leonhard Euler (1707–1783):

$$\pi = 4 \int_0^1 \frac{dx}{\sqrt{1-x^4}} \int_0^1 \frac{x^2 dx}{\sqrt{1-x^4}} \qquad (1.4)$$

or

$$\pi = \lim_{n \to \infty} \frac{4}{n^2} \sum_{k=0}^{n} \sqrt{n^2 - k^2} \qquad (1.5)$$

You can admire a whole range of other π formulae in our collection of formulae on page 223.

Mathematics professors sometimes use yet another definition in their lectures.

Under this definition, $\frac{\pi}{2}$ is declared simply as the *smallest possible zero of the cosine function*. For it is the case that the function $\cos x$ is positive at $x = 0$, but later on, for example when $x = 2$, it is negative. Since in between it runs continuously without any sudden changes, it must cross the x-axis at some point, which is defined as $\frac{\pi}{2}$.

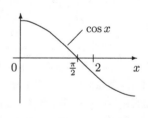

It is difficult to believe, but this definition provided grounds for the important mathematician Edmund Landau, who taught at Göttingen University in Germany, to lose his chair in 1933, since it was stated that such a definition was "un-German". You will find further details of this disgraceful story later on (see page 213).

The number π has fascinated people throughout the ages, but it is unlikely that there have ever been as many π enthusiasts as there are today. This is suggested by the fact that on the Internet world-wide web (WWW) there are at least 200 *home pages* explicitly devoted to π and thousands of online documents referenced by them. It is possible to obtain everything to do with π from the Internet, including some clever things. Several of the statements made in this book you will also find somewhere in the Internet. You will find it is quite different with our CD-ROM, which contains a lot of information you will probably not be able to read anywhere else. It also contains a large number of π-relevant Internet addresses in the file `pilinks.htm` and in the `pilinks` directory.

There are also websites on which renowned π-researchers introduce themselves, for example, Yasumasa Kanada and the brothers Peter and Jonathan Borwein, who developed the algorithms which Kanada used to perform his π calculations. The Borweins and a number of other researchers also make their publications available for reading and downloading.

There are a number of cheerful π clubs on the Internet. It is possible to become a formal member, although in some cases you have to pass an entry test. This test often consists of reciting by heart the first digits of π. For example the "Freunde der Zahl Pi" (Friends of the number Pi) in Vienna, which also publishes a journal called "pi vobiscum", requires successful candidates for membership to be able to recite the first 100 digits by heart[5]. Some aids for remembering π are presented in Chapter 3.

When π is under discussion, often the number e is not far away. This may be because that number, which is also known as the "base of natural logarithms" and whose value 2.71... is fairly close to π, is also a transcendental number. Another reason could be that the two numbers are linked by a wonderful formula by Euler (1.10), to which we shall refer below.

Otherwise, π and e do not have a lot in common. e is a relatively recent discovery, only about 400 years old, whereas π is 10 times as old; e is an offshoot of infinitesimal analysis which was developed in the 17th century when mathematicians learned how to calculate with limits. e is just such a limit. A commercial way of explaining e is to ask to what value an initial sum of money would grow in a year if the rate

---

[5] http://pi314.at/

of interest payable on it was 100% per interest period and this interest period became ever shorter, from a year to half a year, three months etc. until eventually it became "infinitesimally small". The answer to this question is that the end value is e times as great as the initial investment.

The number e has been calculated to far fewer digits than the number π. As of February 1999, 200 million digits had been calculated by Sebastian Wedeniwski, exactly one thousandth of the number of digits to which π has been calculated. (The Chudnovsky brothers are said to have calculated e to 1 billion digits, but this has not been officially confirmed.) Due to this disparity it is assumed that e is in principle more difficult to calculate than π, but this has not been proven.

The latest world record makes π the mathematical constant which has been calculated to the greatest number of decimal places. The inverse value of π, i.e. $\frac{1}{\pi} = 0.31830\,98661\ldots$, has similarly been calculated to 206.1 billion decimal places. Then comes the square root of 2 ($\sqrt{2} = 1.41421\,35623\ldots$), with 137 billion ($\approx 2^{37}$) decimal places calculated. All three calculations have been performed by Kanada. These are followed (as of October 1999) by $e = 2.71828\,18284\ldots$ (as mentioned above, either 1 billion or 200 million), Riemann's number $\zeta(3) = \sum_{n=1}^{\infty} \frac{1}{n^3} = 1.20205\,69031\ldots$ (128 million) and $\ln(2) = 0.69314\,71805\ldots$ (108 million).

A table of such world records can be found on the Internet at http://www.lacim.uquam.ca/pi/records.html.

Not only are the first 206.1 billion decimal places of π known, but individual digits beyond that are also known. Thus, for example, since February 1999 we have known the 10 trillionth digit of π from its hexadecimal representation, an A. This and other such astonishing findings resulted from the use of a new approach to the calculation of π which was discovered quite by surprise in 1995. This subject is discussed in Chapter 10.

Whereas the decimal representation of the 206.1 billion π digits is still awaiting an artistic representation, the formal representation of π is always an art at the same time. No doubt the fact that there are grandiose formulae for π has contributed to its fascination. Here is a selection of the available formulae, arranged in historical order:

1. François Viète (1540–1603) developed the first infinite product for $2/\pi$ in 1593:

$$\frac{2}{\pi} = \frac{\sqrt{2}}{2} \cdot \frac{\sqrt{2+\sqrt{2}}}{2} \cdot \frac{\sqrt{2+\sqrt{2+\sqrt{2}}}}{2} \cdots \qquad (1.6)$$

The formula shows that π can be expressed solely in terms of the number 2 (Compare the formulae (16.137) on page 234).

2. Lord William Brouncker (c. 1620–1684) made his mark in 1658 with the first ever continued fraction for π:

$$\frac{4}{\pi} = 1 + \cfrac{1^2}{2 + \cfrac{3^2}{2 + \cfrac{5^2}{2 + \cfrac{7^2}{2 + \ldots}}}} \qquad (1.7)$$

The unusual feature about this continued fraction is the fact that it contains a regular pattern $(1^2, 3^2, 5^2, \ldots)$, however the price for this is that it is not a *simple* continued fraction as its numerators do not $= 1$. There is also a simple continued fraction for π, but, as mentioned, it does not exhibit any regularity.

3. From 1650 to at least 1973 virtually all calculations of π were performed using arctan formulae. The arctan formula used most frequently was first developed by John Machin (1680–1752), who used it in 1706 to generate what was then a world record of 100 decimal places:

$$\frac{\pi}{4} = 4 \arctan \frac{1}{5} - \arctan \frac{1}{239} \qquad (1.8)$$

4. The next sequence discovered by the Indian mathematician S. Ramanujan (1877–1920) in 1914 has the attraction that it converges extremely quickly. Every term in this series produces 8 accurate digits of π:

$$\frac{1}{\pi} = \frac{\sqrt{8}}{9801} \sum_{n=0}^{\infty} \frac{(4n)!\,(1103 + 26390n)}{(n!)^4 \, 396^{4n}} \qquad (1.9)$$

5. But of all the mathematical formulae, the prize goes to the following theorem by Leonhard Euler dating from 1743. It combines five base quantities π, e, i, 0 and 1) and four basic operators ($+$, $=$, $\cdot$ and exponentiation) into a single expression:

$$e^{i\cdot\pi} + 1 = 0 \tag{1.10}$$

This formula has always held a certain fascination, possibly because it is attractive to look at, although it is not at all easy to understand. What it says is that that if the transcendental number e is first raised to the power of the transcendental number π and then raised again by the imaginary number i, the result is equal to -1.

After the important American mathematician Benjamin Peirce (1809–1880) had proved the formula to his students, he cried: "Gentlemen, that is surely true, it is absolutely paradoxical; we cannot understand it, and we don't know what it means. But we have proved it, and therefore we know it is the truth." [39, p. 585].

In the meantime there is at least one relationship between π and e which comes out without imaginary numbers [73, p. 98]:

$$\frac{e+1}{e-1} = 2 + 4\sum_{r=1}^{\infty}\frac{1}{(2\pi r)^2 + 1} \tag{1.11}$$

This formula establishes a link between the "growth number" e and ever greater circumferences ($2\pi r$).

Again and again the number π has attracted poets and philosophers. In the short story *Contact* by Carl Sagan an extraterrestrial tells an Earth woman that π contains an important message for us in encrypted form. At some point in π, the woman is told, the digits suddenly cease to be randomly distributed numbers and then for a long time only zeros and ones occur, representing the product of 11 prime numbers. It is thus an 11-dimensional message. The sender is thus using mathematics to communicate with humans. His message will even be authenticated, as the zeros and ones only occur in the *decimal representation* of π, which in turn is only meaningful to beings who possess ten fingers. This important message has been in π for trillions of years waiting until 10-fingered mathematicians with fast computers are able to unravel it — which, of course describes present-day man.

This reflection entirely explains the enthusiasm for π all over the world. It also comes at the right time for π researchers, who always need new reasons to justify their research grants.

Unquestionably, π is a "natural constant" which is valid throughout the universe, applying to all living creatures and in the same way.

It would thus seem well suited to the initiation of communication with strange beings. It has therefore been suggested that π should be painted on all the space probes which leave our solar system.

Douglas R. Hofstadter, who caused a sensation a few years ago with three ultra-clever books, presents the following sentence in "Metamagicum".

If π were equal to 3, ...

What Hofstadter means is, firstly, that if π were equal to 3, no one would bother to write the sentence in the first place. Instead, it would be the case that π *is equal to* 3, so the subjunctive would not be necessary. Secondly, all circles would be hexagons, as only hexagons have the perimeter : diameter ratio of 3. And we would probably not use a round o, but a hexagon instead.

Algorithms, Computers and Arithmetic

The history of the calculation of π may be broken down into three distinct phases.

The first era began around 250 BC with the Greek mathematician Archimedes of Syracuse. Whereas up to then only experimental methods had been used, he was the first person to develop a systematic procedure for arriving at limits for π. He calculated the circumferences of regular polygons which he placed in and around a circle. The outer polygons have a perimeter that is greater than the circumference of the circle and therefore provide an upper approximation for π, whereas the inner polygons have a smaller perimeter and therefore produce a lower approximation. The more sides the polygons used have, the smaller the difference between these approximations. Archimedes began with regular hexagons and progressed through 12-, 24- and 48-sided polygons to polygons with 96 sides. Using this method he obtained for π a lower limit of $3\frac{10}{71}$ ($= 3.1408\ldots$) and an upper limit of $3\frac{1}{7}$ ($= 3.1428\ldots$). Both values are accurate to two decimal places, the first being more accurate than the second, which is still more accurate than our decimal 3.14.

From then on, anyone who found two statements too much used the upper limit of $3\frac{1}{7}$ for π, which as a result became possibly the longest-living standard in the world. In the Middle Ages, a few scholars still believed this to be exact. But even more important than the goodness of fit of the two approximations was the fact that Archimedes and his polygon method determined the direction in which virtually all succeeding π mathematicians proceeded for nearly 2000 years. Over the centuries mathematicians increased the number of sides in the polygons used for the calculations until by the end of this era of geometric approximation in 1630 AD, π was known to 39 decimal places.

The second era began in the middle of the 17th century after the discovery of infinitesimal analysis and infinite expressions. A specific method, namely the arctan formulae method (see Chapter 5), dominated π calculations after that for more than 300 years until around 1980. The first extensive calculation of π on this basis produced exactly 100 digits in the year 1706 (using paper and pencil), and the last one in 1973 (using a computer) somewhat more than 1 million.

We are currently in the third era of π calculation. It began around 1980 when mathematicians discovered how to utilise a combination of three independent developments.

The first development was the speeding up of an apparently simple arithmetic operation, namely the multiplication of long numbers. As this is the primary operation involved in calculations of π, speeding it up reduces the total time required by a similar amount. Surprisingly, the new method of multiplication known as *FFT multiplication* has only been around since 1965. Under this method, the time required to perform multiplication operations no longer rises quadratically as the number of factors is increased, as had been the case in what was previously virtually the only multiplication "school", but more or less linearly. In Chapter 11 we discuss this subject in more detail; a numeric example in that chapter indicates that the time required to multiply two factors each of one million digits using the new multiplication procedure is reduced from one day to three seconds.

The second major breakthrough was the development of high-performance algorithms specifically designed for calculating π. They surpass the arctan formulae by several orders of magnitude.

To the extent that these algorithms still utilise infinite series, they are much more productive than even the best arctan series. We have already mentioned such a series produced by Ramanujan with 8 digits per series term (1.9). An even more effective series was developed by the Chudnovsky brothers (8.7), who used it in 1989 to calculate 15 digits per term and thus to attain what was then a world record of 1 billion digits.

Other π algorithms which operate in an entirely different fashion were developed at the beginning of the 1980s primarily by the Canadian brothers, Jonathan (born 1951) and Peter Borwein (born 1953). One of these algorithms is capable of discovering in each step *four times* as many accurate digits as in the previous step. This "quartic" (i.e. gigantic) progression enabled Kanada to calculate his 206.1 billion π digits in 1999 in only 20 iterations. The general design of such procedures is discussed in more detail in Chapters 7 and 9. The German mathematician Carl Friedrich Gauss (1777–1855) had in fact already developed precisely such an ultra-fast algorithm in the 19th century. However, for reasons which are not entirely obvious his procedure was overlooked for over 170 years.

The third development, which is virtually taken for granted these days, is the performance explosion of the computer. For some time now we have observed that the "standard computer" doubles its speed at least every two years. This alone has contributed by a factor of 1000 or more to the explosion in π digits since the beginning of the 1980s. Measured in these terms, the 2 million digits which brought Kanada his first world record in 1981 are the equivalent of the first 2 billion digits in his latest world record in 1999.

With all three developments harnessed in parallel, there has been a huge increase in productivity. Since 1981 the number of known π digits has increased 100,000-fold to the current record of 206.1 billion digits. This is the equivalent of a growth factor of 190% per year. Few other technologies have developed so rapidly.

Why?

What motivates people to pursue those elusive π digits? For practical calculations, 10 or even fewer digits are generally sufficient. 39 digits of π are sufficient to calculate the volume of the universe to the nearest atom. Some scientific applications require intermediate results with significantly more digits than the final result, but we are talking here of no more than a few hundred digits. It is conceivable that studies of mathematical problems using computers could require a few thousand digits, but certainly no more than that [13]. The precision to which π can be calculated far surpasses any practical need, so why continue beyond that?

One practical use for large π programs is to test computer systems. Calculating billions of π digits means executing trillions of arithmetical and logical single operations, so that if there are any hardware errors in a computer subjecting it to such an ordeal is bound to uncover them. In fact there are known instances in which large π programs have brought to light subtle logic errors in computers. Error-free running of a large π program today constitutes an essential ingredient of quality testing new processors.

The number π is the classical and ultimate testbed for numeric analysis. Because so many decimal places are known with certainty, new techniques and methods can be verified using π as a yardstick. Every extra digit that is acquired magnifies this advantage.

The main reason for the preoccupation with π, however, is the unresolved questions which this number raises. What we do *not* know about π is not only enormous, but it is also enormously interesting.

π research offers extraordinary breadth. It expands far into analysis, number theory, function theory, complexity theory, the study of algorithms, statistics and other areas beside. Such a wide spectrum naturally exerts an exceptional fascination.

For 4,000 years π has proved an inexhaustible source of new discoveries and surprises. This has been reaffirmed only in the last few years, which have seen the discovery of novel procedures for the calculation of π, such as the "spigot algorithm" (see Chapter 6) and the "BBP algorithm" (see Chapter 10).

Empirical data can often be an advantage when approaching theoretical issues, and the digits of π provide a wealth of empirical knowledge. Moreover, today we possess aids, such as computer algebra, to enable us to tackle theoretical questions experimentally, and here too it is helpful to have a large stock of digits available.

Every world record achieved is crying out to be bettered. Anyone who has the necessary ability and (access to) resources to achieve something which no one else has yet managed will try. This is true both of sportsmen and scientists.

π mathematics does not end with the development of formulae, theorems and algorithms. The calculation itself is still mathematics, at least calculations beyond the first million digits. Without in-depth knowledge of arithmetic, asymptotic behaviour and Fourier transforms it is not possible to write a π program that will keep up with the competition.

Someone once said that one can turn a normal person into a π fan but that the reverse is impossible. He was quite right. The enthusiasm and pleasure which leaps out of texts written by π digit-hunters suggests that π lives a life of its own in these people. The authors have certainly been caught by the bug.

And last but by no means least, new discoveries about π meet with the interest even of people who are not mathematicians. Put another way, π matters. π is one of those rare objects in mathematics which act as a showpiece. The hunt for ever more π digits provides the opportunity to gain an insight into the research laboratories of mathematicians, who otherwise tend to live out of sight of the public.

New goals

For some years, since around 1995, π mathematicians have increasingly been turning to new goals. "Pi: a 2000-year search changes direction," so Victor Adamchik and Stanley Wagon describe the new trend [2]. Instead of continually calculating more π digits *all over again*, people are now devoting ambition and resources to the calculation of individual digits *at the far end of π*.

The impetus for this new development came from the unexpected discovery of a method which allows individual hexadecimal digits in π to be calculated without having to calculate all the digits which come before. The basis for this is the following new formula:

$$\pi = \sum_{n=0}^{\infty} \frac{1}{16^n} \left(\frac{4}{8n+1} - \frac{2}{8n+4} - \frac{1}{8n+5} - \frac{1}{8n+6} \right) \qquad (1.12)$$

The inventors of this formula (David Bailey, Peter Borwein and Simon Plouffe) won world-wide admiration when they presented it in October 1995 (see Chapter 10). Up to then it was difficult to imagine that individual digits could be resolved independently. A π calculation is like an enormous "tree" of single operations, and the task of calculating a single "leaf" would appear impossible, rather like removing a needle from a haystack without moving the hay on top of it to one side. But with the new formula it is possible, as it were, to fly over π in helicopter and then pick individual locations on which to land. The critical factor here is the magnitude of 16^n in the denominator of all the terms in the series.

When the inventors of the "BBP series" (named after the initial letters of their names) introduced the new theory, they also delivered proof of its efficacy with the announcement of the identity of the 10 billionth hexadecimal digit of π, which at the time was far off from discovery.

The new idea seems to have electrified π researchers all around the world. Since then they have found more formulae of the same type as the BBP series and even an algorithm for generating additional such formulae. It was with one of these new formulae that digits 100 times further on in π than the last π digit calculated by Yasumasa Kanada were successfully calculated.

The most recent and youngest world record holder in *this* π discipline is Colin Percival, who was only born in 1981 and has only recently gone up to university. Unlike Kanada, he did not perform his individual digit calculations on a university supercomputer but divided up the calculations over 1700 computers on the Internet all over the world. He invited owners of computers to help him with his project, called the *PiHex project*, and sent them his calculation program by e-mail, which the computers then executed in their dead time when they did not have anything more important to do. Every computer took on a different subtask and Percival then put the individual pieces together. In this way after two years elapsed time and almost 700 years of computer time (P90 CPU) he attained a new world record on the 11th September 2000: he had found the 250 trillionth hexadecimal digit of π, an E. Because this E consists of the four binary digits 1110, Percival was able to proudly announce:[6]:

The quadrillionth bit of π is 0.

The one slight imperfection of these π calculations is that they only provide *hexadecimal* or binary digits of π, rather than decimal digits. To be precise, to the present date we still do not know a way to calculate individual decimal digits in π which is quicker than calculating all the digits which precede them.

But this problem seems to be only a question of time. At any moment someone could devise a "BBP formula" for calculating individual *decimal* digits of π. We look forward to the day.

---

[6] http://www.cecm.sfu.ca/projects/pihex/announce1q.html

2. How Random is π?

2.1 Probabilities

What is the probability that decimal position s of π will contain the digit z?

At first glance the answer appears quite simple. Because position s must be occupied by one of the 10 decimal digits $0, 1, \ldots, 9$, the probability of any one of them occurring is exactly 10%.

On the other hand, a so-called probability subjectivist, i.e. a follower of the teachings of Thomas Bayes (1702–1761), would say that the answer depends on how much he already knows about π, since according to this philosophy, "probability" measures the extent of what one does not know. If he already knows which number occurs in position s, then the probability that it is a z is either 0% or 100%. If on the other hand he knows nothing about position s, then *as far as he is concerned* the probability that the number will be z is 10%. "No!" mathematicians or followers of classical philosophy will respond, "the basic question is fundamentally flawed." For the positions of π are not random but precisely determined. For example, in the second decimal place there is a 4, so there is no point in asking what the probability is that a 5 will occur in that position. Where certainties reign, there is no need for probabilities.

But if the basic question is now not a reasonable mathematical question, is there another question which we can ask about the "randomness" of π? There is indeed: "Is π normal?"

2.2 Is π normal?

Mathematicians call a decimal number "normal" when every sequence of n digits is equally likely to occur. Thus, for example, in a normal number the digit 0 occurs with a frequency of 1/10 and the sequence of digits 357 with a frequency of 1/1000.

If all ten digits occur equally often in a decimal number, it is said to be "simply normal".

The question as to whether a number is normally distributed is only meaningful and interesting if it is of infinite length. The number π has been proven to be irrational, i.e. it is a number that contains an infinitely long and non-periodic decimal fraction, and therefore it is possible that it could also be normal. But we do not know this. No one has succeeded in proving either that π is normal or that it is not. Nor has anyone been able to demonstrate that such a proof is impossible.

If π were not normal, then this would manifest itself in individual digits or sequences of digits occurring at unequal probabilities. For example, the number 7 might occur more frequently than 3, or perhaps from some point in the sequence there would be no further instances of the combination 314159265. Although there are certainly intervals in which such irregularities (or regularities, depending on how one looks at it) occur, it is not known today whether this is also the case when one views the entire sequence of π.

In a number which is normal, there should always be a point somewhere where, say, the number 5 occurs a million times in succession. But we do not know whether such a sequence also occurs in a number such as π, of which we do not know whether or not it is normal. In a random sequence it easy to find an interval which *apparently* is non-random.

One thing is certain: the digits of π are not random numbers because they can be calculated.

The fact that π is not only irrational but also transcendental does not exclude the possibility of regular patterns occurring within its number sequence. Conversely, the existence of a pattern in the decimal sequence of π does not mean that π is "non-normal". To illustrate this point, let us consider the artificial number $0.1\,2\,3\ldots10\,11\,12\,13\ldots$ which is made by writing all the natural numbers n in sequence after the decimal point. Clearly this number follows a pattern, yet it is normal. This has been proven by Ivan Niven [86].

The preoccupation with the question of whether π is normal has provided the impetus to many statistical studies. The results of these studies are in part instructive, in part disappointing, in part curious. We will now report some of those findings.

2.2 Is π normal?

There are several statistical tests available which enable one to ascertain to what extent the decimal digits in π are randomly arranged; however, the answers are always expressed as statements of probability. If, for example, such a test is used on a particular roulette table, then no result can provide 100% certainty that the table has not been rigged, as even on the "cleanest" table an "impossible" sequence will occur at some point. However, it is possible to quantify the probability of this.

One statistical test, for example, is the "poker test". In the poker game, seven different combinations of five cards are considered "poker hands", e.g. "a pair" or "full house". In the poker test, the actual number of poker hands is compared with the number that one would expect. For example, 2 million "poker hands" occur in the first 10 million decimal digits of π. Their distribution is as follows [117]:

Poker hand	Pattern	Actual number	Expected number
All different	abcde	604,976	604,800
One pair	aabcd	1,007,151	1,008,000
Two pairs	aabcd	216,520	216,000
Three identical	aaabc	144,375	144,000
Full house	aaabb	17,891	18,000
Four identical	aaaab	8,887	9,000
Five identical	aaaaa	200	200

The comparison does not exhibit any striking departures from the numbers one would expect. In fact, the χ^2 test produces the value of 53% for this distribution, which means it is completely normal with no striking features. Only if percentages above 95% or below 5% were obtained, would one view a distribution as "suspect".

If one considers smaller intervals, e.g. intervals of 500,000 decimal places, which are sufficiently large to allow one to expect even the rarest poker hand, "5 identical", to still occur 10 times, the picture is different. For example, the interval from 3,000,001 to 3,500,000 has the following distribution:

Poker hand	Pattern	Actual number	Expected number
All different	abcde	30,297	30,240
One pair	aabcd	50,263	50,400
Two pairs	aabcd	10,877	10,800
Three identical	aaabc	7,156	7,200
Full house	aaabb	927	900
Four identical	aaaab	459	450
Five identical	aaaaa	21	10

In this case, an opponent would very likely suspect that all was no longer as it should be, especially if he examined the number of "Five identical". The probability that this distribution occurs in a random sequence is only 2.6%, so the alarm bells would definitely start ringing. On the other hand, in the next equal-sized interval, the probability reverts once more to 69.5%, and when one considers both intervals together the overall probability is 32.0%, which would certainly not arouse suspicion.

There is also the possibility of the opposite occurring, namely that the distribution of the poker hands matches the expected values so exactly as once again to look suspicious. Such an interval begins, for example, in position 4,250,001. In that sequence the distribution of poker hands is only 0.5% away from the expected figures.

2.3 So is π not normal?

In an early edition of his "Mathematical Games" column Martin Gardner reports a conversation with a "Dr. Matrix" [54]:

"Dr. Matrix borrowed my pencil and rapidly jotted down the value of π to 32 decimals."

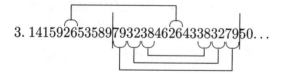

"Mathematicians consider the decimal expansion of π a random series, but to a modern numerologist it's rich with remarkable patterns."

He bracketed the two appearances of 26. "Twenty-six, you observe, is the first two-digit number to repeat. Note how the second 26 marks the center of a bilaterally symmetric series." Dr. Matrix inserted vertical bars to enclose 18 numerals, then bracketed six other number pairs shown in the illustration. "The number pairs 79, 32 and 38 on the left are balanced by the same three pairs in reverse order on the right!" He also called attention to the sets of five digits on each side of the first 26: "The first set sums to 20, the number of decimals preceding the second 26. The second set sums to 30, the number of decimals preceding the second bar. Together they sum to 50, the two-digit number following the last bar. The sequence between bars starts with the 13th decimal and 13 is half of 26. The three pairs – 79, 32 and 38 – have six digits that sum to 32, the pair in the middle as well as the total number of decimals shown. The 46 and 43 on each side of the second 26 sum to 89, the number pair preceding the first bar..."

2.4 The 163 phenomenon

If you were not convinced by Dr. Matrix, then what do you think of this next case?
The number

$$e^{\pi\sqrt{163}} \qquad (2.1)$$

looks particularly contrived. It contains nothing but "unusual" components - the transcendental numbers e and π and the prime number 163, plus a square root and an exponentiation. Anyone would suppose that the combination of such numbers must be an "awkward" number.

But in fact expression (2.1) produces the following remarkable result:

$$e^{\pi\sqrt{163}} = 262,537,412,640,768,743.\underline{99999\,99999\,992}\ldots \qquad (2.2)$$

That is very nearly an integer. The difference between this and an integer is less than 10^{-12}.

This "163 phenomenon" was brought to light by the Scottish mathematician Alexander Aitken (1895–1967) [15, p. 386]. As far as he was concerned it was no accident. In the mathematical world Aitken has become well-known not only for his academic achievements (4 books and 70 publications) but also for his legendary abilities in mental arithmetic. He was able to break down even large numbers into their factors in his head and to say straight off and with unswerving accuracy whether a number called out to him was a prime number or not. One can just hear him saying, "The number 'smells' prime." To elucidate the 163 curiosity one would no doubt need to be cast of the same mould.

One would perhaps first try seeing whether there any other values of n for which $e^{\pi\sqrt{n}}$ is almost integer. In fact there are some. Have a look at the following table:

n	$e^{\pi\sqrt{n}}$
6	2,197.990...
17	422,150.997...
18	61,4551.992...
22	2,508,951.998...
25	6,635,623.9993...
37	199,148,647.99997...
43	884,736,743.9997...
58	24,591,257,751.99999 98...
59	30,197,683,486.993...
67	147,197,952,743.99999 8...
74	545,518,122,089.9991...
163	262,537,412,640,768,743.99999 99999 992...

The author of the list, Roy Williams, offered a prize for anyone who could either convincingly demonstrate that this is a chance occurrence or else explain to an intelligent university graduate why it is not a chance occurrence.

One thing is certain: the length of the list alone suggests that not all of these instances are chance. The list is actually not even complete, as there are many other values of n above 163 for which $e^{\pi\sqrt{n}}$ produces three or more nines after the decimal point, for example, $n = 232$(5 nines), 719(4), 1169(4), 1467(8), 4075(5), 5773(4).

But knowing this does not make us any the wiser. For the mathematical proofs are so complicated that, however highly one rates university graduates, the prospects of winning the prize are extremely slim because the phenomenon defies simple explanation. The topic belongs to the theory of modular equations, of which only a few experts have a full grasp [61].

For $n = 43$, 67 and 163 there is the explanation that here the "j function"

$$j(n) = \frac{1}{q} + 744 + 196884q + 21493760q^2 + 864299970q^3 + \ldots \quad (2.3)$$

with $\frac{1}{q} = -e^{\pi\sqrt{n}}$

is an integer and furthermore a perfect cube. For $n = 43$ it has the value -960^3, for $n = 67$ the value -5280^3 and for $n = 163$ the value -640320^3. Why that is so is understood only by specialists. But if we assume that it does behave like this, then in these cases the phenomenon can be explained from the fact that in the series expansion (2.3), $e^{\pi\sqrt{n}}$ differs from an integer by the sum of the terms q, q^2 etc., which are very small and can be neglected.

At any rate, if one knows these facts, then the relationship (2.2) can be written in even more spectacular fashion as follows:

$$\sqrt[3]{e^{\pi\sqrt{163}} - 744} = 640319.99999\,99999\,99999\,99999\,99993\ldots \quad (2.4)$$

There are a few other values of n for which approximating series expansions can be used by way of explanation, even if they cannot explain the j function. The Indian mathematician S. Ramanujan (1887–1920), of whom we shall hear more in Chapter 8, discovered some such series. However, he was not given to lengthy explanations, and his essay [96] is no exception. As far as one can tell, he followed a semi-heuristic path which involved combing through the first few hundred degrees of modular equations. In the course of so doing he came across the degrees 22, 37 and 58, which in a certain combination actually produce integers. For example, he found the relationship $e^{\pi\sqrt{37}} = 64((6+\sqrt{37})^6 + (6-\sqrt{37})^6) - 24 - 4372e^{-\pi\sqrt{37}} - \ldots$, whose approximation then "explains" the case of $n = 37$ in the list if one leaves out the terms after the -24.

With these six values of n, those n which are bigger by a quadratic factor are then also explained, e.g. $n = 232 = 58 \cdot 2^2$ or $n = 1467 = 163 \cdot 3^2$. But we are still no closer to explaining the (many) other cases, e.g. the spectacular case of $e^{\pi\sqrt{719}}$ which is only 0.000013 away from an integer.

The assumption that combination of many complex variables inevitably produces a complex result is deceptive, as we have seen. This was also the experience of Donald E. Knuth of Stanford University, California. Knuth, whom we shall encounter several times more in this book, is the author of the multi-volume work *The Art of Computer Programming*, which is one of the most cited (and possibly even most read) books in the field of computer science. In the introduction to the second volume, Knuth tells the story of his super-random number generator [77, p. 4], with which he planned to generate the most random of all random numbers.

To this end Knuth wrote a program which was designed to be especially *random*. For example, on every pass it branched off to a *random* place in the program and for each random number it performed a *random* number of loop passes. Knuth also coded the program in such a complicated way that no one could understand it. Then he started with an initial value, which naturally was chosen *at random*. He expected that since his program contained so much randomness it would produce incredibly random numbers.

But what happened? After starting up, the program converged almost immediately on the 10-digit number 6065038420, which thereafter transformed itself back into itself in only 27 cycles. When he tried a different initial value it took 3178 cycles, but even this value may be regarded as lamentable.

The conclusion he drew was that one should never generate random numbers with a random method.

This applies to investigations into π as well. Thus, we should add that not only is the 163 phenomenon no chance occurrence but it even constitutes the background for one of the fastest methods of calculating π. It was on the basis of the underlying theory that the Chudnovsky brothers developed their brilliant "Chudnovsky series" (8.7) which enabled them in 1989 to break through the one billion decimal places of π mark for the first time

2.5 Other statistical results

Naturally the digits of π have been counted by computers in many different ways. For example, when Yasumasa Kanada issued his statement about his 206.1 billion decimal place world record in October 1999, he published along with it the distribution of the first 200 billion digits of π. It goes like this:

2.5 Other statistical results

Number	Occurrences
0	20,000,030,841
1	19,999,914,711
2	20,000,136,978
3	20,000,069,393
4	19,999,921,691
5	19,999,917,053
6	19,999,881,515
7	19,999,967,594
8	20,000,291,044
9	19,999,869,180
Total	200,000,000,000

As you see, the numbers are very evenly distributed. The number which deviates most is the number 8, but even this number deviates by only 0.00001% from the mean. The χ^2 test produces a probability of around 55% for the randomness of this distribution, i.e. "statistically insignificant".

Kanada also searched through his record-breaking π for interesting patterns. He reports, for example, that the sequence **01234567891** occurs five times among the 206.1 billion decimal places of π, which is somewhat striking as one would only expect this sequence to occur twice. Another peculiarity is the fact that the sequence **543210987654** appears for the first time relatively late in π, at the end of the 206 billion, in position 197,954,994,289, although there is a 50% probability that this sequence would appear in the first 69 billion decimal places. This probability figure is obtained from inserting $w = 0.5$ and $k = 11$ in the formula below (2.5), which states that the probability of a k-digit number not appearing in a number containing n digits is

$$n = \frac{\log w}{\log(1 - 10^{-k})} \tag{2.5}$$

$$w = (1 - 10^{-k})^n \tag{2.6}$$

According to the inverse formula (2.6), the probability of the fact ascertained by Kanada (despite the inherent difficulty of making such a statement) is only 13%. Even so, such numbers would not arouse the suspicion of any statistician.

If one divides up the digits of π into blocks of 10 decimals, what is then the probability that any given block will consist of all different numbers? One such block could, for example, be the sequence 0123456789.

There are three points to make here.

Firstly, the probability of this case is greater than one might think. We are looking for 10! hits out of a total 10^{10} possibilities. This fraction can be calculated easily with a pocket calculator, and the result is the astonishingly large value of around $0.036\% = 1/2755\ldots$. In other words, a block of 10 different figures will occur on average as frequently as every 2755 blocks.

Secondly, in reality this case actually occurs much earlier than is suggested by this probability, which is already quite large. In fact a hit occurs as early as the seventh block, as you can see from this list of the first seven blocks of 10 digits:

$$\begin{array}{l} 3.1415926535 \\ 8979323846 \\ 2643383279 \\ 5028841971 \\ 6939937510 \\ 5820974944 \\ 5923078164 \quad \mathtt{<<<<} \end{array}$$

Thirdly, this result is independent of whether the blocks begin before or after the decimal point of π, because the same number 4 appears immediately before as well as at the end of the block.

2.6 The Intuitionists and π

Kanada's calculation feat has an interesting historical association, as was pointed out by Jonathan Borwein [29].

At the beginning of the 20th century the Dutch mathematician L.E.J. Brouwer developed a theory which bears the name "Intuitionism". In short, Brouwer questioned "the Law of the Excluded Middle", which had been sacrosanct to mathematicians from the time of Aristotle: A statement is either true or false, "tertium non datur" – there is no third possibility. Brouwer argued that this law cannot be applied generally. By way of proof he asked, for example, whether it is true or false that the sequence 123456789 occurs in π. Given the state of π research in his day, he could comfortably assume that this question would never be answered, since to do this, one would "probably" have to know 6.9 billion decimal places of π, whereas at the time (1908) only 707 decimal places were actually known. So if one cannot say whether this statement is true or false then, according to Brouwer, no one can invoke the Law of the Excluded Middle for this statement.

Today, that question *has* been answered (thanks to Kanada), and with it a number of examples which Brouwer and his disciples used to prove their theory have been rendered of no value. One of these examples was the following.

One of the implications of tertium non datur is that proof of the impossibility of an impossible characteristic is also a proof of the characteristic itself. Brouwer's counterarguments went as follows. I write down the sequence of decimal digits for π and underneath it the decimal fraction $\rho = 0.33333\ldots$, which I terminate as soon as the sequence 0123456789 occurs in π. If the 9 in the sequence is the kth decimal place, then $\rho = (10^k - 1)/(3 \cdot 10^k)$.

Now let us assume that ρ is irrational. If that is so, then ρ cannot be equal to $(10^k-1)/(3 \cdot 10^k)$, as that would be a fraction and consequently a rational number. Hence the stated number sequence should also not occur in π. But if it does not occur in π, then $\rho = 1/3$, which again is a fraction, contrary to the hypothesis.

In this way, the assumption that ρ is not a rational number leads to a contradiction. Despite this we cannot assert that ρ is rational, as that would mean that we would have to be able to state two integer numbers p and q for the numerator and denominator of ρ, and to do that we would either have to locate a sequence 0123456789 in π or else to prove that there is no such sequence.

This and other examples were thus used by the Intuitionists to support their thesis that the Law of the Excluded Middle could not be assumed to be always correct. Not only the example cited above but many other examples were based on the assumption that we would never know for certain whether the number sequence 0123456789 occurs in π or not.

Unfortunately for the theory, we do now know, and so the question arises of whether this refutes the theory. In fact it does not. The proponents of the theory would only have had to choose a much longer sequence in their examples, and they can still do that even today, for example they could suggest a sequence consisting of 100 billion zeros in π, and everything would be back to the way it was before. Not only is Intuitionism not refuted, but in a certain way it is even confirmed,

2.7 Representation of continued fractions

The decimal sequence of π reveals little that would enable us to draw conclusions as to the non-randomness of the π digits. So are there perhaps other representations of π which are more informative? This question is actually quite sensible, for every alternative representation of π, such as the hexadecimal representation, is based on a quite different calculation tree, and there is every likelihood that the new variant of π could look quite different.

A promising approach here is the so-called simple continued fraction of π[1]. As already mentioned on page 1, the "related" transcendental number e appears quite differently when represented as a continued fraction compared with its decimal representation. Whereas in its decimal form e begins with $2.71828\,18284\,590\ldots$, expressed as a continued fraction the number begins like this:

$$e = 2 + \frac{1}{1+}\frac{1}{2+}\frac{1}{1+}\frac{1}{1+}\frac{1}{4+}\frac{1}{1+}\frac{1}{1+}\frac{1}{6+}\frac{1}{1+}\frac{1}{1+}\frac{1}{8+}\frac{1}{1+}\cdots \quad (2.7)$$

The decimal representation of e thus possesses no regularity, whereas its representation as a continued fraction exhibits a lot of regularity: the formula is made up of sequences of three fractions whose denominators are of type $1\,n\,1$, in which every n is always 2 higher than the n in the previous sequence of three terms.

The number π does not exhibit any such characteristics. In the case of π, the continued fraction (apparently) does not follow any construction rule:

$$\pi = 3 + \frac{1}{7+}\frac{1}{15+}\frac{1}{1+}\frac{1}{292+}\frac{1}{1+}\frac{1}{1+}\frac{1}{1+}\frac{1}{2+}\frac{1}{1+}\frac{1}{3+}\frac{1}{1+}\frac{1}{14+}\cdots$$

One can perhaps conclude from this that π is structured even more randomly than e.

In 1985 William R. Gosper calculated 17 million continued fraction positions of π, so it is possible to perform statistical analyses of these as well.

The results for the first 8192 elements (3, 7, 15 etc.) include the following:

---

[1] We go into continued fractions in more detail in Section 4.4 on pages 64.

Mean value of 1st 2048 elements	27.5
Mean value of 2nd 2048 elements	15.3
Mean value of 3rd 2048 elements	8.2
Mean value of last 2048 elements	10.4

Is there anything in this? The mean value of the first 2048 elements appears to be quite out of line. It is almost twice as big as the mean value of all the elements and it is also significantly greater than the mean of the two next equally large sections. The picture does not change if one splits the number sequence up into smaller sections or divides it in different ways: every time the first or at least the foremost mean values are greater than those which come afterwards.

When one examines the data more closely, however, even this discovery turns out not to amount to much. The fact is that quite early on, at position 431 of the continued fraction expansion of π, an unusually large element occurs, 20776, which has the effect of distorting the statistics. In the first section of 2048 elements shown above, that single element by itself accounts for 10.14 of the mean value. If one ignores the influence of that element, then this first section becomes inconspicuous.

It is unfortunately the case that the simple continued fraction of π, like the decimal (and every other) representation of π, contains no other message than that there is little regularity in the irregularity of the number.

However, there are other continued fraction representations than the simple continued fraction and in some of them there is a certain amount of regularity. One example here is the elegant new continued fraction for π which was developed by L.J. Lange in 1999:

$$\pi = 3 + \frac{1^2}{6} + \frac{3^2}{6} + \frac{5^2}{6} + \frac{7^2}{6} + \cdots \qquad (2.8)$$

Lange, 1999 [80]

Here, all the elements follow the simple construction rule $\frac{(2n-1)^2}{6}$.

The known π digits pass all random number tests with flying colours. For this reason one can use π as a random number generator – though one must not reveal to "the opponent" where the random numbers came from. For a lottery company it would not be a good idea to simply select the next 7 digits of π as the hits each week.

So is there at least a ray of hope that a sensation might yet be uncovered in the digits of π? Perhaps. The above-mentioned Chudnovsky brothers, who have made a significant contribution to π research and are above the suspicion of starting rumours simply in order to create a sensation, hinted a few years ago (1992) that they might have found something: "It's unfortunately not statistically significant yet (but) it's close to the edge of significance." They told the interviewer, Richard Preston, that they would need many billions of digits in order to be able to say anything more precise [92]. It seems that what the brothers were hinting at was a property which eludes common statistical tests, namely possible "waves" in the decimal sequence. If, for example, the first, third and fifth billions *regularly* have fewer than the expected number of a particular digit and the second, fourth and sixth billions *regularly* have more than the expected number of that digit, the χ^2 test would probably not detect this, but it would still provide major cause to reflect.

Since April 1999 the Chudnovsky brothers now have their "many billions of digits", courtesy of their arch-rival in the world record stakes. They now have the evidence they need to either confirm or reject their hypothesis. We suspect that it will be a case of the latter, but we are eagerly awaiting the results.

3. Shortcuts to π

Unquestionably the easiest way to arrive at π is to follow the cover of this book, which shows the first few dozen decimal places. 5000 decimal places are listed on page 240f and 400 million decimal places are contained on our CD-ROM. Another easy way to access π is with a computer algebra system such as *Mathematica* or a special π program, of which there are several on the CD-ROM and dozens more on the Internet.

However, this chapter concentrates on approaches to π which involve the exercise of a certain amount of initiative.

3.1 Obscurer approaches to π

If area A and diameter d of a circle are known, then π is equal to $4A/d^2$. This is precisely the approach which underlies the following curious C program created by Brian Westley. It calculates π to 4 decimal places.

```
#define _ 00>00?0:--00,--F;
int F,00;
main(){F_00();printf("%1.3f\n",-4.*F/00/00);}F_00()
{
                        _ _ _ _
                  _ _ _ _ _ _ _ _ _
                _ _ _ _ _ _ _ _ _ _ _
              _ _ _ _ _ _ _ _ _ _ _ _ _
            _ _ _ _ _ _ _ _ _ _ _ _ _ _ _
          _ _ _ _ _ _ _ _ _ _ _ _ _ _ _ _
          _ _ _ _ _ _ _ _ _ _ _ _ _ _ _ _
        _ _ _ _ _ _ _ _ _ _ _ _ _ _ _ _ _
        _ _ _ _ _ _ _ _ _ _ _ _ _ _ _ _ _
        _ _ _ _ _ _ _ _ _ _ _ _ _ _ _ _ _
        _ _ _ _ _ _ _ _ _ _ _ _ _ _ _ _ _
        _ _ _ _ _ _ _ _ _ _ _ _ _ _ _ _ _
          _ _ _ _ _ _ _ _ _ _ _ _ _ _ _ _
          _ _ _ _ _ _ _ _ _ _ _ _ _ _ _ _
            _ _ _ _ _ _ _ _ _ _ _ _ _ _
              _ _ _ _ _ _ _ _ _ _ _ _ _
                _ _ _ _ _ _ _ _ _ _ _
                  _ _ _ _ _ _ _ _ _
                        _ _ _ _
}
```

The "circle" here has an area of 201 (measured using the characters _ and -_) and a diameter of 16 (measured in lines made out of these characters), so that a value of 3.141 is obtained for "π". To obtain more decimal places, you need only to make the circle larger, each additional decimal place requiring the size of the circle to be increased by a factor of around 10.

The program does not look at all like an ordinary C program, but it even complies with the ANSI C programming standard. With this program, Westley took a prize at the *International Obfuscated C Code Contest* (IOCCC) in 1988. This competition has taken place every year since 1984 on the Internet. The winner is the person who creates the most incomprehensible and most creative C program, which, however, must be executable. The basic tools used are the concise C syntax and above all the C pre-processor (as here). The results are usually artificial yet also artistic. But they are invariably examples of how one should not write a program.

Here is another example of such programming:

```
                                                                char
                                                    _3141592654[3141
          ],__3141[3141];_314159[31415],_3141[31415];main(){register char*
       _3_141,*_3_1415, *_3__1415; register int _314,_31415,__31415,*_31,
        _3_14159,__3_1415;*_3141592654=__31415=2,_3141592654[0][_3141592654
      -1]=1[__3141]=5;__3_1415=1;do{_3_14159=_314=0,__31415++;for( _31415
     =0;_31415<(3,14-4)*__31415;_31415++)_31415[_3141]=_314159[_31415]= -
    1;_3141[*_314159=_3_14159]=_314;_3_141=_3141592654+__3_1415;_3_1415=
    __3_1415      +__3141;for                  (_31415 = 3141-
           __3_1415    ;                       _31415;_31415--
            ,_3_141 ++,                        _3_1415++){_314
           +=_314<<2 ;                         _314<<=1;_314+=
           *_3_1415;_31                        =_314159+_314;
           if(!(*_31+1)                        )* _31 =_314 /
           __31415,__314                       [_3141]=_314 %
           __31415 ;* (                        _3__1415=_3_141
           )+= *_3_1415                        = *_31;while(*
           _3__1415 >=                         31415/3141 ) *
           _3__1415+= -                        10,(*--_3__1415
          )++;_314=_314                        [_3141]; if ( !
          _3_14159 && *                        _3_1415)_3_14159
          =1,__3_1415 =                        3141-_31415;}if(
          _314+(__31415                        >>1)>=__31415 )
          while ( ++ *                         _3_141==3141/314
          )*_3_141--=0                         ;}while(_3_14159
          ) ; { char *                         __3_14= "3.1415";
          write((3,1),                         (--*__3_14,__3_14
         ),(_3_14159                           ++,++_3_14159))+
         3.1415926; }                          for (__31415 = 1;
         _31415<3141-                          1;_31415++)write(
         31415% 314-(                          3,14),_3141592654[
         _31415    ] +                         "0123456789","314"
         [ 3]+1)-_314;                         puts((*_3141592654=0
        ,_3141592654))                         ;_314= *"3.141592";}
```

This work of C–art created by Roemer B. Lievaart looks as if it is all about the number π, especially since it produces 3141 decimal digits. Ironically enough, it actually calculates *not* π but its cousin, the number e. Its output therefore begins with 2.7128....

We will explain elsewhere which algorithm was used in the program.

3.2 Small is beautiful

Next we present a mini program only 133 characters long, once again programmed in ANSI C, which we believe is the shortest C program in the world for the calculation of π – naturally only until someone outdoes us.

```
a[52514],b,c=52514,d,e,f=1e4,g,h;
main(){for(;b=c-=14;h=printf("%04d",e+d/f))
for(e=d%=f;g=--b*2;d/=g)d=d*b+f*(h?a[b]:f/5),a[b]=d%--g;}
```

Despite its short length, this little program calculates 15,000 decimal places of π.

The underlying calculation method is the so-called spigot algorithm, under which the program produces a steady trickle of digits one at a time without using or requiring previously computed digits. You can see the method in action for yourself if you run our JAVA applet `spigot/pispigot.htm` which is on our CD-ROM. The spigot algorithm and the little program are discussed in more detail in Chapter 6 from page 77.

This algorithm was discovered by Stanley Rabinowitz, who published it in 1991 in the form of a program written in FORTRAN [94]. The author provided virtually no explanation or proof, instead promising his readers all would be explained later on in a major article. This appeared a mere four years later in 1995 [95] (with Stanley Wagon as co-author). One can therefore say that programming of the spigot algorithm preceded its rationale.

Our version shown above evolved in a number of stages, in which several authors (Dik T. Winter, Achim Flammenkamp and others) were involved.

3.3 Squeezing π through a sieve

π can be approximated using a method which recalls the "sieve of Eratosthenes" devised by the Greek mathematician of that name (284-202 BC) as a means of collecting prime numbers.

Beginning with the sequence of natural numbers $1, 2, 3, 4, \ldots$, one first of all removes every second element, starting from the third element:

$$1, 2, 4, 6, 8, 10, 12, 14, 16, 18, 20, 24, 26, 28, 30, 32, 34, \ldots$$

Starting from the fifth element, every third element is now removed:

$$1, 2, 4, 6, 10, 12, 16, 18, 22, 24, 28, 30, 34, \ldots$$

then, from the seventh element, every fourth element:

$$1, 2, 4, 6, 10, 12, 18, 22, 24, 30, 34, \ldots$$

and so on. As a general rule, on the kth pass, every $(k+1)$th element is removed from the $(2k+1)$th element.

The procedure produces the sequence of values $f(n) = \{1, 2, 4, 6, 10, 12, 18, 22, 30, 34, \ldots\}$. When these values are sequentially entered in the formula

$$\frac{n^2}{f(n)} \tag{3.1}$$

each result is a closer approximation to π. For $n = 1, 2, 4, 8, 16, 32, \ldots 4096 = 2^{12}$, the following approximations are obtained:

$$1, 2, 2.7, 2.9, 3.12, 3.10, 3.08, 3.12, 3.1405, 3.13, 3.1423, 3.1414, 3.1412$$

Unfortunately the increase in accuracy is only modest. The series converges only of the order of $O(n^{4/3})$, which is a little better than linear convergence. Furthermore, the memory requirements are considerable: to use $f(4096)$, as in the present case, for the approximation, one needs a sieve with 5.34 million holes in it.

K.S. Brown discovered that the sequence of $f(n)$ can also be obtained by intelligent *rounding up* . His algorithm goes as follows.

Take an integer n and round it to the next multiple of $n - 1$, then to the next multiple of $n - 2$, and so on until the next multiple of 1. The result is then exactly the same $f(n)$ as one obtains from the sieve procedure. For example, with $n = 10$, one obtains the intermediate values 18, 24, 28, 30, 30, 32, 33, and 34, with the end result $f(10) = 34$.

3.4 π and chance (Monte Carlo methods)

The needle problem of the Comte de Buffon

During the American Civil War, Captain C.O. Fox was recovering from a wound in a military hospital. To pass the time, he threw a number of identical needles *in random fashion* onto a board on which he had previously drawn a series of parallel lines each a needle's length apart. He counted the number of throws and the number of hits, i.e. instances in which a needle touched or intersected a line.

After 1100 throws, the Captain had determined π to two decimal places. How come?

First of all it seems to have been the Comte de Buffon (1707–1788), who examined this kind of experiment and in whose honour it is now known as the *Buffon needle problem*. In 1777, Buffon showed that the ratio of hits to throws was $2{:}\pi$, or, stated otherwise, that the *probability* of a needle thrown at random onto the area coming to rest across one of the lines was, $\frac{2}{\pi} \approx 63.7\%$. With this knowledge, Fox was able to calculate π by doubling the number of throws and dividing by the number of hits.

The interesting thing about the needle problem is that it forges a link between the "geometric" π and the quite different area of probabilities. There are other similar relationships between π and chance, from which other methods of calculating π are derived. They are informatively called *Monte Carlo methods*.

The dartboard algorithm

Imagine a circle with radius $r = 1$ with a circumscribed square of side length $2r = 2$. "Darts" are thrown at this square, following a random distribution, whereby a "hit" is deemed to occur every time the dart lands in the circle. The procedure is roughly similar to the standard game of darts, except that in this case the "black" is much larger than in a conventional dartboard.

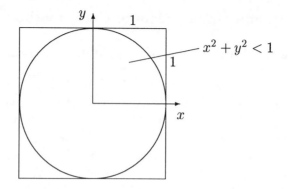

The procedure is simplified by only considering the first quadrant.

As the number of throws (n) increases, the number of hits (t) approaches the ratio of the area of the quadrant of the circle to the quadrant of the board.

$$\lim_{n\to\infty} \frac{t}{n} = \frac{A(\text{circle quadrant})}{A(\text{board quadrant})} = \frac{\pi r^2/4}{r^2} = \frac{\pi}{4} \qquad (3.2)$$

$$\pi \approx \frac{4t}{n} \qquad (3.3)$$

The result is a neat algorithm for approximating π. You only need to simulate a number of throws (as many as possible) and test on each throw whether the dart lands inside the circle. This test is particularly simple: if x and y are the co-ordinates of the landing point($0 \leq x, y \leq 1$), then the point will be inside the circle precisely when its distance from the origin is $\sqrt{x^2 + y^2} < 1$, i.e. $x^2 + y^2 < 1$.

In the following C++ program the random co-ordinates x and y are generated using the standard C++ function rand(); as this function delivers (pseudo-) random numbers between 0 and (the system-dependent variable) RAND_MAX, they must be divided again by RAND_MAX to obtain co-ordinates in the interval from 0 to 1. The random number generator is initialised through the standard function srand() with a time-dependent, and hence varying, value:

```
// Dartboard-Algorithm for approximating pi

#include <iostream.h>
#include <stdlib.h>
#include <time.h>
#include <math.h>
```

```
int main(void)
{
    long    k, n, hits;
    const double factor = 1.0 / RAND_MAX;

    while (1)
    {
        cout << "Enter the no of tosses (or 0 to exit): ";
        cin >> n;
        if ( n <= 0) // input <= 0 means end-of-job
            break;
        // Initialize the random generator
        srand((int)clock());

        // Throw n tosses
        for (k=hits=0; k < n; ++k)
        {
            // Find two random numbers within 0..1
            double x = rand() * factor;
            double y = rand() * factor;
            if (x*x + y*y < 1.0) // Within circle ?
                ++hits;          // yes: hits += 1
        }
        double pi_approx = 4.0 * hits / n;
        cout    << "Approximation of pi after "
                << n << " tosses: " << pi_approx
                << " (error="
                << fabs(M_PI - pi_approx)*100/M_PI
                << "%)\n";
    }
    return 0;
}
```

π and coprimality

The probability of two integer numbers chosen at random being relatively prime to each other is $6/\pi^2$. This characteristic can be used for another π approximation procedure, whereby a sufficiently large number of pairs of random numbers is calculated and these are examined for coprimality. From the ratio of the number of coprime pairs t to the total number of trials n the following approximation for π is obtained:

$$\lim_{n \to \infty} \frac{t}{n} = \frac{6}{\pi^2} \tag{3.4}$$

$$\pi \approx \sqrt{\frac{6n}{t}} \tag{3.5}$$

Ball [14] reports a field experiment in which 50 students were asked to each write down five pairs of random numbers. It turned out that 154 of them were coprime. This led to $6/\pi^2 = 154/250$, and hence to $\pi = 3.12$.

Instead of playing this game with students, one should try it with a computer. The only problem with such a program is how to investigate the results for coprimality. Fortunately, for 2300 years there has been an algorithm which does this and could be described as the "grandfather" of all algorithms. It was first noted down by the Greek mathematician Euclid (ca. 325–265 BC) and appears in Book VII of his 13-volume *Elements*. This work is the most successful mathematics book ever written and has been published more than 1000 times since antiquity.

The *Euclidean algorithm* can be written down in only three lines of source code (e.g. see the program listing below), although it takes longer to explain in words. Starting from two integer numbers $A > 0$ and $B > 0$, it calculates the largest integer which divides into both A and B without a remainder. If this "highest common factor" (HCF) $= 1$, then A and B are coprime to each other. The algorithm utilises the property that $A = \lfloor \frac{A}{B} \rfloor B + C$, with $0 \leq C < B$. For example, let us consider the integers $A = 78$ and $B = 21$. As $78 = 3 \cdot 21 + 15$, the highest number which divides into both 78 and 21 is also the largest number which divides into 21 and 15. Through iteration one obtains:

$$78 = 3 \cdot 21 + 15$$
$$21 = 1 \cdot 15 + 6$$
$$15 = 2 \cdot 6 + 3$$
$$6 = 2 \cdot 3 + 0$$

Hence 3 is the highest common factor of 78 and 21.

In the following little program we simulate the student experiment and so arrive at an approximation for π.

```
// Approximation for pi by a Monte-Carlo-Method:
//
// The program finds pairs of random integers.  It determines
// whether they are relatively prime. The theoretical probability
// of these events is 6 / pi^2.

#include <iostream.h>
#include <stdlib.h>
#include <math.h>

int euclid(int u, int v)
{
    int r;
    while ( (r = u % v) != 0 )
    {
```

3.4 π and chance (Monte Carlo methods)

```
      u     = v;
      v     = r;
   }
   return v;
}

int main(void)
{
   while (1)
   {
      int     n, nTries, nHits;

      cout << "Enter no of tries (or 0 to exit): ";
      cin >> n;
      if (n <= 0) break;
      srand((int)clock());
      for (nTries=nHits=0; nTries < n; ++nTries)
      {
         int A = rand() + 1;
         int C = rand() + 1;
         if (euclid(A, C) == 1) // A and C are relative prim
            ++nHits;
      }
      double f  = nHits * 1.0 / nTries;
      double pi = sqrt(6.0 / f);
      cout << "After " << nTries
           << " tries is pi " << pi << endl;
   }
   return 0;
}
```

The following approximations have been obtained through running the programs a number of times:

Number of trials	Approximation for π	Error
10	3.464102	+0.322509...
100	3.273268	+0.131675...
1,000	3.194383	+0.052790...
10,000	3.142438	+0.000845...
100,000	3.143913	+0.002320...
1,000,000	3.141554	−0.000038...

Although Monte Carlo methods are attractive and interesting and they are easy to implement, they are not very well suited to calculations of π, as they converge poorly. Even after a million attempts it is scarcely feasible to obtain more than four correct decimal places of the π. And it is perfectly possible to obtain a poorer approximation of π on a larger number of trials than on a smaller, as can be seen from the table above. Methods which are based on random numbers are probably the least efficient of all systematic procedures for the calculation of π.

On top of this there are two further fundamental problems. Firstly, computer programs are not capable of generating truly random numbers as there is inevitably a point at which they reproduce numbers they have already generated, in the same sequence. Secondly, the "pseudo-" random numbers generated by a computer inevitably lie between certain limits, so that on this basis alone they are not truly random. It follows from this that Monte Carlo methods do not strictly speaking converge on π but, rather, oscillate around π. However, given the poor convergence obtained with these methods, these are really only esoteric thoughts.

3.5 Memorabilia

Another way of arriving at decimal places of π is to learn them by heart.

A number of *mnemonic verses* are available to assist with this. We mentioned already such a verse in the first chapter. Many people gifted at playing with words have tried to conjure up texts in which the number of letters in each successive word corresponds to the corresponding digit of π.

An English poem intended to teach π to 31 decimal places was submitted by a person "F. S. R." to the scientific journal *Nature* in 1905:

```
Sir,   I   send   a   rhyme   excelling
 3     1    4     1     5         9
In   sacred   truth   and   rigid   spelling
 2      6       5      3      5        8
Numerical   sprites   elucidate
    9          7           9
For   me   the   lecture's   dull   weight
 3     2    3        8        4       6
If   Nature   gain   not   you   complain
 2      6      4      3     3       8
Tho   Dr   Johnson   fulminate.
 3     2      7          9
```

Probably easier to memorize is the following text, written by the English astrophysicist Sir James Jeans (1877–1946), which produces 24 digits:

How I want a drink, alcoholic of course, after the heavy
chapters involving quantum mechanics. All of thy geometry,
Herr Planck, is fairly hard...

Such texts are available in many languages. The Internet address
http://users.hol.gr/~xpolakis/PiPhilology.htm mentions more
than 60 poems in 24 languages, among them "$14+X$" in English and
"$4+X+V$" in German. (Some of them seem harder to remember than
the native digits of π.)

Soon after the Frenchman Fautet de Lagny (1660–1734) succeeded
in 1717 in calculating π to 127 decimal places, P. Decerf tried his
hand at turning this into a work of art and composed a monumental
π poem 127 words long. Some years later it was discovered that de
Lagny's "π" contained a transcription error, so that the 112th decimal
digit should have been an 8 rather than a 7. It was then necessary
to change Decerf's poem and publish an "update", something which
rarely happens in the world of poetry [41, p. 153].

Perhaps the longest mnemonic poem in the world was composed by
Michael Keith. He has written a "Cadaeic cadenza" which produces
an awe-inspiring 3865 decimal places of π[1]. Cadae is a made-up word
consisting of the letters which correspond to the first five π digits
3 1 4 1 5.

The first section of the cadenza is a modified version of Edgar Allen
Poe's famous poem of 1844 *The Raven*. The new work was given the
ambiguous title of *Near A Raven*. It consists of 18 verses and begins
as follows:

---

[1] http://users.aol.com/s6sj7gt/cadenza.htm

Poe, E.   3.1
Near A Raven   415

Midnights so dreary, tired and weary.	926535
Silently pondering volumes extolling all by-now obsolete lore.	897932384
During my rather long nap – the weirdest tap!	62643383
An ominous vibrating sound disturbing my chamber's antedoor.	27950288
"This", I whispered quietly, "I ignore".	419716
Perfectly, the intellect remembers: the ghostly fires, a glittering ember	939937510
Inflamed by lightning's outbursts,	8209
windows cast penumbras upon this floor.	749445
Sorrowful, as one mistreated, unhappy thoughts I heeded:	92307816
That inimitable lesson in elegance – Lenore –	406286
Is delighting, exciting ... nevermore.	2089

...

The title and the first two verses alone produce 80 digits of π. These are followed by a further 16 such verses plus the author's signature for the other 660 places. Keith worked hard to remain as close to the original as possible as regards the metre, action, melody and rhythm. Thus for example he succeeded in retaining the melodious dark refrain at the end of every verse.

Of the 14 sections of the Cadaeic cadenza, section 3 contains another poem by the mathematician and π enthusiast Charles Lutwidge Dodgson, better known as Lewis Carroll, author of "Alice in Wonderland", which has likewise been adapted to π, while section 11 even contains a work by William Shakespeare (who we all know wrote the famous words, "to π or not to π").

π texts reach their first critical point at the 32nd decimal place, when the first zero appears in the decimal sequence of π. As there are no words with zero letters, the authors of mnemonic verses have to think up something special. Keith's solution to the problem is to use words of 10 letters, such as, for example, "disturbing". Other authors require their readers to interpret a punctuation mark as a 0. A particularly difficult problem here is the 601st decimal place of π, where a sequence of three zeros occurs in succession.

Short after the 740 decimal places of "Near The Raven", in section 2 of his work, Keith had to resolve another tricky decimal sequence in π, the so-called Feynman point, for at position 762 there is a devilish sequence of 7 numbers 9999998! This sequence may be beautiful to π aesthetes, but for poor Mike, who had to find a text containing a sequence of six 9-letter words and one 8-letter word in succession, it was bad news. Nevertheless, he succeeded in cracking it.

There are a number of methods of memory training which have evolved from the example of π. These include: the use of recurring patterns, creation of blocks of fixed or variable length, association of graphic, rhythmic, musical or colour representations, learning by repetition at varying intervals and situations. There are articles available on this subject on the Internet.

Anyone who knows a lot of digits of π will go far. For example such a person can take part in a π recitation competition or become a member of a π club, as these normally require one to be able to recite a number of digits of π by way of entry test. We have already mentioned the Austrian club "Freunde der Zahl Pi" (Friends of the number Pi) which requires of candidates seeking admission that they should be able to recite 100 decimal places in a public place in front of a π notary. The Swedish "1000 Club"[2] even requires 1000 digits of π.

The members of these clubs are not just nut-cases who fill their heads with digits of π, but serious scientists seem to take pleasure in being able to perform this feat. One of these is Simon Plouffe, who was one of the three mathematicians who recently developed a novel π algorithm, the BBP algorithm (see page 117). Plouffe obtained an entry in the (French) Guinness Book of Records in 1977 by memorising 4096 digits of π. In fact, he said, he actually knew 4400 digits by heart, but 4096, i.e. 2^{12}, was a nicer number.

Since 1995 the world record holder in the recitation of π digits has been Hiryuku Goto, then aged 21, who recited 42,000 digits in nine hours. Since this new record was established it has been claimed that Japanese is better suited for memorising sequences of numbers than other languages. Or perhaps Goto simply had nothing more sensible to do after programmers had eliminated him from their code ...

3.6 Bit for bit

The inverse value of π, $1/\pi$, may be represented either as $= 0.3183\ldots$ or $= 0.01010\,00101\,11110\ldots$. This is no contradiction, as the second

---
[2] http://www.ts.umu.se/~olleg/pi/club_1000.html

value is just the binary representation, in which only zeros and ones occur.

To obtain the places after the decimal point in the binary representation of $1/\pi$ from the decimal version, keep multiplying the decimal fraction by 2 and writing each calculated position in front of the point as the next binary digit while keeping the fraction for the next multiplication. Thus $0.3183 \times 2 = 0.6366$, hence 0 is the first binary digit. $0.6366 \times 2 = 1.2732$, so the second binary digit is 1. Continuing with the post-decimal point result of the last operation, i.e. with $0.2732 \times 2 = 0.5464$, the third binary digit is another 0. And so the procedure continues. Of course the precision of binary representation is not any higher than that of the decimal representation, so that from n accurate decimal places one can obtain at the most $1/\log_{10} 2 \approx 3.3$ times as many accurate binary digits.

A quite different method for calculating the binary representation of $1/\pi$ was discovered by Simon Plouffe of Bordeaux University. His procedure was proved and generalised in 1995 by Jonathan Borwein and Roland Girgensohn [36].

In Plouffe's bit recursion method the starting point is $a_0 = \tan(1) = 1.5574\ldots$. From this, succeeding values of $a_1, a_2, a_3, \ldots$ are calculated using the following rule:

$$a_{k+1} = \frac{2a_k}{1 - a_k^2} \qquad (3.6)$$

Each a_k is only checked to see whether it is negative or not. If it is negative, then the next binary digit is a 1, otherwise it is a 0.

The first 10 a_k and the resulting binary decimal places run as follows:

k	a_k	Binary position of $1/\pi$
0	+1.5574	0
1	−2.1850	1
2	+1.1578	0
3	−6.7997	1
4	+0.3006	0
5	+0.6610	0
6	+2.3478	0
7	−1.0406	1
8	+25.111	0
9	−0.0797	1

Interesting and unexpected as this algorithm is, it conceals a trap. Namely, before you can calculate $1/\pi$ in this way, you first need the initial value $\tan(1)$, but this takes longer to calculate than the direct calculation of $1/\pi$.

3.7 Refinements

You might already have a few π digits and be wondering whether you could build on them. You can. An elegant method for doing this was developed by Daniel Shanks[3] [108].

If p_0 is an approximation for π which has n accurate digits, then the following calculation step produces an improved approximation p_1, which is accurate to three times as many digits:

$$p_1 = p_0 + \sin p_0 \tag{3.7}$$

Example: with $p_0 = 3.14$, i.e. $n = 3$, $p_1 = p_0 + \sin p_0$ produces the new approximation of $p_1 = 3.14159\,265\ldots$. The new value is accurate to 9 decimal places.

The derivation is simple if one uses the series formula for $\sin \varepsilon$ and remembers that $\sin(\pi + x) = -\sin x$. In the following, let ε be the error of the previous value p_0:

$$p_0 = \pi + \varepsilon$$
$$p_1 = p_0 + \sin p_0$$
$$= \pi + \varepsilon + \sin(\pi + \varepsilon)$$
$$= \pi + \varepsilon - \sin \varepsilon$$
$$= \pi + \varepsilon - \left(\frac{\varepsilon}{1!} - \frac{\varepsilon^3}{3!} + \frac{\varepsilon^5}{5!} - \cdots\right)$$
$$= \pi + \frac{\varepsilon^3}{6} - \cdots$$

With n accurate decimal places in the initial value, p_0 comes to $\varepsilon < 10^{-n}$. Hence, the accuracy of the new approximation is $\varepsilon^3/6 < 0.2 \cdot 10^{-3n}$, i.e. at least $3n$ decimal places.

Shanks shows that it is possible to achieve still more accurate approximations in one step. Thus,

---

[3] In 1961 Shanks together with Wrench passed the 100,000 mark in a π calculation [106].

$$p_2 = p_0 + \frac{2\sin p_0 - \tan p_0}{3} \tag{3.8}$$

$$= \pi + \frac{\varepsilon^5}{20} + \cdots \tag{3.9}$$

improves the approximation p_0 of π by more than five times as many accurate places. Shanks may have learned this trick from Wildebrod Snell (1581–1626) (see page 183).

3.8 The π room in Paris

The only museum in the world to possess a π room is the *Palais de la Découverte* in Paris, Av. Franklin Roosevelt. You will find that the museum does not have a house number as it is very big and forms part of the *Grand Palais* building complex.

At the entrance we follow the sign *salle pi* and a really good article from the *Mathematical Intelligencer* [67]. They will both take us to room 31 [sic!], which is round. There we can learn a few (mainly French) facts about π and marvel at the first 707 digits of π which are displayed in three spiral revolutions around our heads.

These 707 decimal places were calculated in 1874 by William Shanks (see page 195). Unfortunately he made a mistake at the 528th position which remained undetected for some time. Thus, at the international exhibition "Art and technology in modern life" in 1937, it was in the π room of all places that an incorrect π was displayed [48, p. 105].

When the error was discovered by Ferguson in 1945, the museum corrected its sequence of numbers immediately.

But without success. The rumour that there is an error in the π on show there persists even to this day. Thus the 1997 edition of the French encyclopaedia *Quid* contains a photograph of the π room as it appears today, i.e. with the correct π, but with the caption,

"In the dome of the Palais de la Découverte in Paris the first 627 [sic] decimal places of the transcendental number π are displayed, but apparently some of them are incorrect."

Some people just cannot accept the truth!

4. Approximations for π and Continued Fractions

4.1 Rational approximations

The shortest approximation for π is simply 3. This is 4.5% away from the true value of π and in fact this approximation occurs twice in the Bible (see page 169). The longest approximation is 206.1 billion digits long and is still not entirely accurate. Only when we write "π" for π[1], are we entirely accurate - everything else is longer and either more or less of an approximation.

For the purpose of simplification in this chapter we shall use $\pi(s)$ to designate a π approximation to s correct decimal places. The approximation 3 used in the Bible we will thus designate as $\pi(0)$. In Babylon the approximation $\pi(1) = 3\frac{1}{8} = 3.125$ was used more than 1000 years earlier.

The statement that $\pi(s)$ is a π approximation to s correct decimal places means that it is less than 10^{-s} out from the true value, $|\pi - \pi(s)| < 10^{-s}$. On the other hand, the converse conclusion does not hold true: an absolute error of $< 10^{-s}$ does not always also guarantee s correct decimal places. For example, the good approximation 3.1416 differs from $\pi = 3.14159265\ldots$ by only $0.000007\ldots$, which is less than 10^{-5}, but it is still only accurate to three decimal places. However, this particular problem can only occur when the approximation is greater than π.

In the old days before infinite decimal fractions were known and understood, it was suggested that one might approximate a "ratio" such as the ratio of the circumference of a circle to its diameter using a ratio, i.e. using a fraction consisting of a numerator and a denominator. The classic π approximation of this kind is that developed by Archimedes around 250 BC:

---

[1] or one of the many mathematical identities (see the collection which starts on page 223).

$$\pi(2) = \frac{223}{71} < \pi < \frac{22}{7} = \pi(2) \tag{4.1}$$

The fact that this approximation states an interval makes it particularly attractive, although the manner in which it was discovered has its drawbacks. (For more on this, see page 171.) The notation used to specify the interval makes clear that π does not exactly equal either the left-hand or the right-hand limit, whereas representations of π have not always been so honest. Thus, for example, Tsu Chhung-Chih, who discovered the excellent approximation

$$\pi(6) = \frac{355}{113} \tag{4.2}$$

in China c. 480 AD, was convinced that this fraction was *exactly equal* to π.

An approximation is called "rational" if it can be represented by a fraction consisting of integer numerators and denominators. Such representations are extremely common, and all decimal, hexadecimal, binary etc. digit sequences of π are rational approximations since, for example, 3.14 is the fraction created by dividing 314 by 100. Because there is a convention that with certain bases such as 10, 16 or 2 we only "think" the denominator and do not write it, these approximations are concise and effective. In actual fact, however, the approximation $\pi(2) = 3.14$ means not just 4 keystrokes consisting of a decimal point and 3 numerals, but 7 keystrokes consisting of the fraction line and two sets of 3 numerals in each of the numerator and denominator. But if one is going to use the 7-keystroke version, there are other fractions, e.g. the above-mentioned fraction $355/113 = \pi(6)$, which represent π more accurately, in this case 4 decimal places more accurately.

A good rational approximation for π is one in which the ratio of the number of correct decimal places to the number of digits in the numerator and denominator is particularly high. As we shall see shortly, the best rational approximations in this respect attain the value 1.

The best source for good approximations of a (transcendental or irrational) number is its so-called continued fraction. Under this remarkable, but very interesting representation, the denominators are not, as in the decimal representation, powers of a fixed base number but a combination consisting of an integer and a fraction.

In the case of π, the continued fraction begins as follows:

$$\pi = 3 + \cfrac{1}{7 + \cfrac{1}{15 + \cfrac{1}{1 + \cfrac{1}{292 + \cfrac{1}{1 + \cdots}}}}} \qquad (4.3)$$

Alternatively, a tabular representation of its elements looks like this[2]:

b_i	.0	.1	.2	.3	.4	.5	.6	.7	.8	.9
0.	3	7	15	1	292	1	1	1	2	1
1.	3	1	14	2	1	1	2	2	2	2
2.	1	84	2	1	1	15	3	13	1	4
3.	2	6	6	99	1	2	2	6	3	5

In order to arrive at a rational approximation for π all one has to do is terminate one's simple continued fraction at any point and work it out. (Incidentally, one never needs to do any cancelling of fractions to their lowest terms.) For example, after terminating the continued fraction after the 3, 7, 15 and 1, one obtains the approximations 3, $\frac{22}{7}$, $\frac{333}{106}$ and $\frac{355}{113}$, which are accurate to 0, 2, 4 and 6 decimal places respectively.

Of the approximations obtained in this way, those obtained by terminating the continued fraction directly *before* one of the large elements are especially accurate. When one examines the relative error in relation to the terminal element, one obtains a curve like this:

---

[2] The first 2000 elements will be found on page 244ff.

4. Approximations for π and Continued Fractions

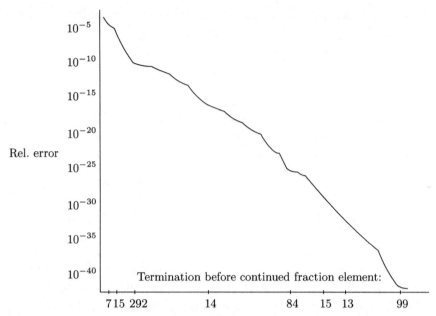

It can be seen from the graph that if termination occurs before a large element such as 15, 292, 84 or 99, the relative error drops off especially sharply, so that excellent approximations are to be found at these points:

Termination point		Approximation
Before	$b_2=15$	$\pi(2) = \frac{22}{7}$
Before	$b_4=292$	$\pi(6) = \frac{355}{113}$
Before	$b_{12}=14$	$\pi(12) = \frac{5\,419\,351}{1\,725\,033}$
Before	$b_{21}=84$	$\pi(21) = \frac{21\,053\,343\,141}{6\,701\,487\,259}$
Before	$b_{25}=15$	$\pi(25) = \frac{8\,958\,937\,768\,937}{2\,851\,718\,461\,558}$
Before	$b_{27}=13$	$\pi(29) = \frac{428\,224\,593\,349\,304}{136\,308\,121\,570\,117}$
Before	$b_{33}=99$	$\pi(37) = \frac{2\,646\,693\,125\,139\,304\,345}{842\,468\,587\,426\,513\,207}$

In favourable cases, the number of accurate decimal places is almost equal to the sum of the number of digits in the numerator and denominator. If one terminates the continued fraction before $b_4 = 292$, $b_{21} = 84$ or $b_{33} = 99$ (the biggest elements), the number of accurate decimal places is exactly equal to the number of digits. There are more such cases further on, e.g. before $b_{77} = 16$, $b_{79} = 161$ or $b_{80} = 45$.

On the other hand, termination before $b_5 = 1$ produces the value $\frac{103993}{33102} = \pi(9)$, i.e. 2 decimal places fewer than the number of digits in the numerator and denominator.

If A_n and B_n are, respectively, the numerator and denominator of a continued fraction for a number κ which is terminated before the element b_{n+1}, then the relative error in the approximation is

$$|(\frac{A_n}{B_n} - \kappa)|/\kappa < \frac{1}{A_n \cdot B_n \cdot b_{n+1}} \tag{4.4}$$

This simple relationship was brought to our attention by F.L. Bauer, and it verifies most effectively the statement that it "pays" especially well to terminate the continued fraction before a large element, since this large element then goes in the denominator of the error estimate, where it has the effect of significantly reducing the value.

No better, i.e. shorter, approximations will be found for the stated numbers of accurate decimal places. This can be proved from the theory of continued fractions. For example, this remarkable rational approximation discovered by Johann Heinrich Lambert (1728–1777)

$$\pi(25) = \frac{1\,019\,514\,486\,099\,146}{324\,521\,540\,032\,945} \tag{4.5}$$

requires an extra 5 keystrokes compared with the equally accurate continued fraction approximation stated in the 5th row of the above table.

The table includes several approximations for π which have made history. The origins of 22/7 (Archimedes) and 355/113 (Tsu Chhung-Chih) have already been mentioned. The two approximations formed on termination before the elements 14 and 13 were discovered in Japan in 1766. What is of especial note here is that at that time continued fractions were still unknown, so that these discoveries must have been made by a different route. How this happened is apparently not known, and hence our wonderment is all the greater.

4.2 Other approximations

However accurate rational approximations may be, they neither look pretty nor are easily remembered. Those qualities are achieved in other, visually pleasing, "artistic" approximations with impressive symbols or fine symmetries.

The Greek philosopher Plato (427–348 BC) is said to have known this approximation [51, p. 126]:

$$\pi(2) = \sqrt{2} + \sqrt{3} \tag{4.6}$$

and his fellow philosopher in India, Zhang Heng (78–139 AD), was the first to work with the formula

$$\pi(1) = \sqrt{10} \tag{4.7}$$

The following approximation is attributed to the author of the "Divine Comedy", Dante Alighieri (1265–1321), who was highly educated in mathematical matters:

$$\pi(3) = 3 + \frac{\sqrt{2}}{10} \tag{4.8}$$

The authors have been unable to discover who invented the following impressive approximation formula, but it may have been the Indian astronomer Aryabhata, who was born in India in 476 AD.

$$\pi(4) = 512\sqrt{2 - \sqrt{2 + \sqrt{2 + \sqrt{2 + \sqrt{2 + \sqrt{2 + \sqrt{2 + \sqrt{2 + \sqrt{2}}}}}}}}} \tag{4.9}$$

One might be tempted to think that the first minus sign was an error, but that is not the case. As one can easily check with elementary geometry, the expression $8\sqrt{2 - \sqrt{2}}$ is the perimeter of an octagon which is inscribed in a circle of radius 1, i.e. with the circumference 2π. Every time the number of sides is doubled, the innermost $\sqrt{2}$ is replaced by the expression $\sqrt{2 + \sqrt{2}}$ and the factor in front of the outermost root is doubled. In this way, the above approximation specifies half the perimeter of an inscribed regular 1024-sided polygon.

Due to the large number of 2s under the roots, the formula (4.9) resembles François Viète's product of 1593 (1.6). In fact the two formulae are related and converge in similar fashion. They both approach the number π through inscribed 2^n-sided polygons, in the case of Viète's product through its area and in the case of the above formula through its perimeter.

Kochansky (1631–1700) discovered

$$\pi(4) = \sqrt{\frac{40}{3} - \sqrt{12}} \tag{4.10}$$

while Carl Friedrich Gauss (1777–1855) was only 14 when he calculated [82, p. 8]:

$$\pi(13) = \frac{22}{7} \cdot \frac{2484}{2485} \cdot \frac{12983009}{12983008} \tag{4.11}$$

The young Friedrich started from the known approximation $\frac{22}{7}$ and then divided it into 3.14159265.... He subtracted the quotient from 1 and converted the remainder to a fraction with the numerator 1. The denominator is around 2485. He then proceeded in similar fashion with $\frac{22}{7} \cdot \frac{2484}{2485}$.

Adrien-Marie Legendre (1752–1833) referred to this approximation [75]:

$$\pi(9) = \ln\frac{1}{x} - 2x^4 \quad \text{where} \quad x = \frac{1}{2}(2^{\frac{1}{4}} - 1)/(2^{\frac{1}{4}} + 1) \tag{4.12}$$

Srinivasa Ramanujan (1887–1920) was a past master at discovering approximations. You will find further information about this important Indian mathematician who was greatly enamoured of π in Chapter 8. In his article entitled *Modular Equations and Approximations to π* [96] of 1914, he presented many approximation formulae. To introduce the idea, the essay starts with the simpler approximations, beginning with an elegant, symmetrical expression [96, p. 34–35]:

$$\pi(3) = \frac{9}{5} + \sqrt{\frac{9}{5}} \tag{4.13}$$

$$\pi(3) = \frac{19}{16}\sqrt{7} \tag{4.14}$$

$$\pi(3) = \frac{7}{3}\left(1 + \frac{\sqrt{3}}{5}\right) \tag{4.15}$$

$$\pi(6) = \frac{99}{80}\left(\frac{7}{7 - 3\sqrt{2}}\right) \tag{4.16}$$

$$\pi(9) = \frac{63}{25}\left(\frac{17 + 15\sqrt{5}}{7 + 15\sqrt{5}}\right) \tag{4.17}$$

These formulae Ramanujan derived from modular equations. They are followed in the same article by a "curious" approximation which he discovered "empirically" [96, p. 35]

$$\pi(8) = \sqrt[4]{9^2 + \frac{19^2}{22}} \qquad (4.18)$$

Attempts have been made to find out what may have been concealed behind the word "empirically". The most likely explanation is the following [25, p. 655]. Ramanujan was an expert in continued fractions, and so he probably noticed that the simple continued fraction of π^4 possesses a particularly striking term:

$$\pi^4 = 97 + \frac{1}{2+} \frac{1}{2+} \frac{1}{3+} \frac{1}{1+} \frac{1}{16539+} \frac{1}{1+} \ldots \qquad (4.19)$$

This observation would have suggested to him that he should terminate the continued fraction before the unusually large element 16539 and use it to find $\pi^4 \approx 97 + \frac{9}{22} = 9^2 + \frac{19^2}{22}$.

Ramanujan obtained another empirical approximation by improving the approximation 355/113. From this he found, "simply from creating the inverse value of $1 - (113\pi/355)$",

$$\pi(14) = \frac{355}{113}\left(1 - \frac{0.0003}{3533}\right) \qquad (4.20)$$

In 1913 Ramanujan published a one-page article [97] containing a geometric construction for the approximation $\frac{355}{113}$. The essay itself contains nothing remarkable, but what is remarkable, however, is the fact that this paper bore the title *Squaring the Circle* and neither in the title nor in the text did the expression appear with any quotation marks. Since Lindemann had proved in 1882 that it was impossible to square the circle, a fact of which Ramanujan was undoubtedly aware, no mathematician could really afford such a title any more. It is a testimony of Ramanujan's mathematical self-assurance that he did so.

As far as Ramanujan's real field of π approximation was concerned, the previous approximations were only finger exercises. In fact he came up with some considerably more accurate formulae [96, p. 31]:

$$\pi(15) = \frac{24}{\sqrt{142}} \ln\left(\sqrt{\frac{10 + 11\sqrt{2}}{4}} + \sqrt{\frac{10 + 7\sqrt{2}}{4}}\right) \qquad (4.21)$$

$$\pi(18) = \frac{12}{\sqrt{190}} \ln\left((2\sqrt{2} + \sqrt{10})(3 + \sqrt{10})\right) \qquad (4.22)$$

$$\pi(22) = \frac{12}{\sqrt{310}} \ln\left[\frac{1}{4}(3 + \sqrt{5})(2 + \sqrt{2})\left((5 + 2\sqrt{10})+\right.\right.$$

$$+ \sqrt{61 + 20\sqrt{10}}\bigg)\bigg] \tag{4.23}$$

And to crown it all,

$$\pi(31) = \frac{4}{\sqrt{522}} \ln\left[\left(\frac{5+\sqrt{29}}{\sqrt{2}}\right)^3 (5\sqrt{29} + 11\sqrt{6}) \times \right.$$

$$\left. \times \left(\sqrt{\frac{9+3\sqrt{6}}{4}} + \sqrt{\frac{5+3\sqrt{6}}{4}}\right)^6\right] \tag{4.24}$$

These approximations too stem from modular equations and are a proof of Ramanujan's supreme mastery of this kind of equation and function.

More recently, the Borwein brothers have derived further approximations of the type Ramanujan developed [32, p. 194]:

$$\pi(2) = \frac{3(3\sqrt{13} + 7)}{17} \tag{4.25}$$

$$\pi(5) = \frac{103\sqrt{13} + 125}{158} \tag{4.26}$$

$$\pi(7) = \frac{66\sqrt{2}}{33\sqrt{29} - 148} \tag{4.27}$$

$$\pi(8) = \frac{4}{\sqrt{58}} \ln(396) \tag{4.28}$$

$$\pi(9) = \frac{180 + 52\sqrt{3}}{45\sqrt{93} + 39\sqrt{31} - 201\sqrt{3} - 217} \tag{4.29}$$

$$\pi(9) = \frac{12}{\sqrt{58}} \ln\left(\frac{\sqrt{29} + 5}{\sqrt{2}}\right) \tag{4.30}$$

Dario Castellanos is another π enthusiast, who in 1988 in an academic paper on "The Ubiquitous π" [41, p. 79] presented all kinds of folklore and attractive approximations he had developed himself.

He started by transforming the Ramanujan approximation shown at (4.18) into an even more attractive form:

$$\pi(8) = \sqrt[4]{102 - \frac{2222}{22^2}} \tag{4.31}$$

He then tried something new. While continuing Ramanujan's procedure to the fifth power, he discovered

$$\pi(8) = \sqrt[5]{\frac{77729}{254}} \tag{4.32}$$

and was surprised that Ramanujan had not come across this himself. In addition, Castellanos also discovered these further approximations [41, p. 79–80, 83]:

$$\pi(6) = \frac{47^3 + 20^3}{30^3} - 1 \tag{4.33}$$

$$\pi(6) = 1.09999901 \cdot 1.19999911 \cdot 1.39999931 \cdot 1.69999961 \tag{4.34}$$

$$\pi(7) = 2 + \sqrt{1 + \left(\frac{413}{750}\right)^2} \tag{4.35}$$

$$\pi(10) = \left(95 + \frac{93^4 + 34^4 + 17^4 + 88}{75^4}\right)^{1/4} \tag{4.36}$$

$$\pi(11) = \frac{1700^3 + 82^3 - 10^3 - 9^3 - 6^3 - 3^3}{69^5} \tag{4.37}$$

$$\pi(13) = \left(100 - \frac{2125^3 + 214^3 + 30^3 + 37^2}{82^5}\right)^{1/4} \tag{4.38}$$

Simon Plouffe, whom we have already encountered several times in this book, was indefatigable in his search for approximations. He has every reason to be pleased with this one:

$$\pi(6) = \left(\frac{689}{396}\right) / \ln\left(\frac{689}{396}\right) \tag{4.39}$$

or with

$$\pi(8) = \ln(5280) / \sqrt{\frac{67}{9}} \tag{4.40}$$

This next approximation creates a link between π and the "Golden ratio" $\phi = \frac{\sqrt{5}+1}{2}$:

$$\pi(3) = \frac{6}{5}\phi^2 \tag{4.41}$$

while e and π are linked as follows:

$$\pi(2) = \frac{9-e}{2} \tag{4.42}$$

$$\pi(3) = \sqrt[7]{2e^3 + e^8} \tag{4.43}$$

$$e(3) = \sqrt[7]{20 + \pi} \tag{4.44}$$

$$e(7) = \sqrt[6]{\pi^4 + \pi^5} \tag{4.45}$$

These approximations can be transformed into even more impressive forms. Thus, the last approximation (4.45) becomes the attractive and easy-to-remember expression

$$\pi^4 + \pi^5 \approx e^6 \tag{4.46}$$

in which a remarkable 7 digits (403.4287) are identical on both the left-and right-hand side.

Another link between π and e is made in the approximation formula developed by James Stirling (1692–1770) for the calculation of $n!$, i.e. the product $1 \cdot 2 \cdot 3 \cdot \ldots \cdot n$:

$$n! \approx \left(\frac{n}{e}\right)^n \sqrt{2\pi n} \tag{4.47}$$
<small>Stirling, 1730</small>

Quite apart from the good quality of the approximation — the relative error is less than 1% from $n \geq 9$ and less than 0.1% from $n \geq 84$ — this formula once again gives us cause to wonder at the apparent ubiquity of π. Here it occurs in a problem comprising only integers.

In our view, Stirling's formula, due to its beauty, its usefulness and its age, belongs in the "eternal" list of the greatest formulae which features formulae such as Euler's theorem (1.10), mentioned previously.

Incidentally, the formula can be improved significantly through a small correction resulting from the asymptotic expansion:

$$n! \approx \left(\frac{n}{e}\right)^n \sqrt{2\pi \cdot (n + \frac{1}{6})} \tag{4.48}$$

The relative error is now less than 1% from $n \geq 3$ and less than 0.1% from $n \geq 9$.

For the similar case of the product of all the *odd* numbers $< 2n$, F.L. Bauer [17, p. 49] used the following approximation:

$$(2n-1)! = (2n-1) \cdot (2n-3) \cdot \ldots \cdot 3 \cdot 1 \approx \frac{\sqrt{(2n)!}}{\sqrt[4]{\pi \cdot (n + \frac{1}{4})}} \tag{4.49}$$

This approximation follows from Bauer's good π formula (16.62), which converges on π significantly better than the Wallis formula or its derivative (16.61) even though the only difference between the two is a tiny $\frac{1}{4}$ in the denominator.

The π approximations which Daniel Shanks derived from investigating complex quartic number fields are an order of magnitude better. His pièce de résistance is the following astonishingly simple formula which produces π to 80 decimal places [107, p. 398]:

With

$$D := \frac{1}{2}(1071 + 184\sqrt{34}) \tag{4.50}$$

$$E := \frac{1}{2}(1533 + 266\sqrt{34}) \tag{4.51}$$

$$F := 429 + 304\sqrt{2} \tag{4.52}$$

$$G := \frac{1}{2}(627 + 442\sqrt{2}) \tag{4.53}$$

followed by

$$d = D + \sqrt{D^2 - 1} \tag{4.54}$$

$$e = E + \sqrt{E^2 - 1} \tag{4.55}$$

$$f = F + \sqrt{F^2 - 1} \tag{4.56}$$

$$g = G + \sqrt{G^2 - 1} \tag{4.57}$$

we obtain

$$\pi(80) = \frac{6}{\sqrt{3502}} \ln(2 \cdot d \cdot e \cdot f \cdot g) \tag{4.58}$$

In his article, Shanks proves the existence of an even better approximation of this type, which produces as many as 109 correct decimal places for π. But due to the time that this would take he did not calculate them. But, as he puts it, "It could be done."

Of course, a large number of good approximations for π can be obtained from terminating infinite π series at an appropriate point. For example, the Ramanujan series (1.9) shown on page 13, which yields 8 correct decimal places per term, produces an 80-digit approximation of π if it is terminated after the 10th term.

This method might appear somewhat trivial. On the other hand, an infinite series which *almost* produces π and thus converges close to π would not be at all trivial. However many terms one were to calculate in such a series, their sum would *never* come to π.

In fact there are some such series and they are truly amazing. Of the two series shown below, the first one produces π to over 18,000 decimal places and the second one to as many as 42 billion.

$$\pi(18,000) = \frac{\ln 10}{100^2} \left(\sum_{n=-\infty}^{+\infty} \frac{1}{10^{(n/100)^2}} \right)^2 \qquad (4.59)$$

J. and P. Borwein, 1992 [34]

$$\pi(42 \text{ billion}) = \frac{1}{10^{10}} \left(\sum_{n=-\infty}^{+\infty} e^{-(n^2/10^{10})} \right)^2 \qquad (4.60)$$

J. and P. Borwein, 1992 [34]

These series thus converge to numbers which – even theoretically – do not equal π but which "happen to" agree with π in the first 18,000 or 42 billion decimal places, as the case may be.

The Borwein brothers derived these formulae from so-called modular identities, and in fact the path from there to the formulae is surprisingly short [34]. When one considers that they evolved by chance rather than being derived from the underlying theory they are quite remarkable and could only too easily deceive one into thinking that here indeed was the answer to the question of life, the universe and π. The Borweins in fact regard the series as an example of "caveat computat".

Incidentally, these formulae can be turned into any number of better approximations for π by replacing the values 100 or 10^{10} in them with larger powers of 10.

Unfortunately, the series are not suitable for real calculations of π. It is true that they can be easily transformed from the range $-\infty \ldots +\infty$ to the range $0 \ldots +\infty$, and the terms in the first series are derived from simple decimal shifts. However, one needs the value $\ln 10$ in the first formula, and this is even more time-consuming to calculate than π itself, and in the second series one would need to calculate around 30 billion places of e^x.

4.3 Youthful approximations

In the *Jugend forscht* ("youth researches") competition held every year in Germany, Sven Kabus took a first prize in 1998 in Schleswig-Holstein with a π program that was based on the "Aryabhata" approximation (4.9). Kabus [70] first of all ascertained the formula analogous to (4.9) for the perimeter of *circumscribed* 8-, 16-, 32-, ...-sided polygons and observed that (4.9) turned up in it, e.g. in the 16-sided polygon

$$U_{U3} = \frac{2U_{E3}}{\sqrt{2+\sqrt{2+\sqrt{2}}}} \tag{4.61}$$

where U_{Un} and U_{En} refer to the perimeters of circumscribed and inscribed polygons with 2^{n+1} sides. Kabus then made the numerically interesting observation that the perimeters of the circumscribed polygons were always twice as far from the circumference of the circle as those of the inscribed 2^n-sided polygons. He proved

$$\lim_{n\to\infty} \frac{\frac{1}{2}U_{Un} - \pi}{\pi - \frac{1}{2}U_{En}} = 2 \tag{4.62}$$

From the two observations, Kabus then derived the following iterative algorithm for π:

Algorithm (4.63) Kabus (alg 4.63)

Initialise:
$$a_0 := \sqrt{\tfrac{1}{2} + \tfrac{1}{4}\sqrt{2}}$$
Iterate: $(n = 0, 1, 2, \ldots K-1)$
$$a_{n+1} := \sqrt{\tfrac{1}{2} + \tfrac{1}{2}a_n}$$
Then:
$$p_K = \frac{2^{K+2}}{3} \frac{\sqrt{2-2a_K} + \sqrt{2-2a_{K+1}}}{\sqrt{2+2a_{K+1}}} \xrightarrow{1} \pi$$

Each iteration improves the π approximation by 1.2 places, whereas, for example, Archimedes's method only produced 0.6 places each time the number of sides of the polygon was doubled. Adding various cunning improvements in the arithmetic and using a PASCAL program, Kabus arrived in this way at 5000 decimal places of π.

4.4 On continued fractions

Continued fractions are mentioned a number of times in this book and we would therefore like to say a few words about them at this point.

Continued fractions are the "lost sons" of mathematics classroom teaching. They are regarded as too advanced for secondary schools and too elementary for universities, as a result of which they generally fall between the cracks of both syllabuses [18, p. 129]. Their historical roots go back to the 17th century mathematicians Cataldi, Wallis and Huygens, and their theory traces its origins to Leonhard Euler, who presented them in 1748 in his *Introductio in analysin infinitorum* [52, pages 303 ff]. The classic textbook on this interesting topic was written by Oskar Perron [90], and a more recent textbook is that by C. D. Olds [87].

A continued fraction is a fraction whose numerator is an integer and whose denominator is the sum of an integer and a fraction which in turn possesses the same form.

$$b_0 + \cfrac{a_1}{b_1 + \cfrac{a_2}{b_2 + \cfrac{a_3}{b_3 + \cfrac{\ddots}{\quad\cfrac{a_{n-1}}{b_{n-1} + \cfrac{a_n}{b_n}}}}}} \tag{4.64}$$

$$= b_0 + a_1/(b_1 + a_2/(b_2 + a_3/(b_3 + \cdots + a_n/b_n))) \tag{4.65}$$

This representation, which rapidly grows to occupy a large physical area, is often replaced by

$$b_0 + \frac{a_1|}{|b_1} + \frac{a_2|}{|b_2} + \frac{a_3|}{|b_3} + \cdots + \frac{a_n|}{|b_n} \tag{4.66}$$

<small>A. Pringsheim, 1898</small>

or by

$$b_0 + \frac{a_1}{b_1} + \frac{a_2}{b_2} + \frac{a_3}{b_3} + \cdots \frac{a_n}{b_n} \tag{4.67}$$

<small>L.J. Rogers, 1907</small>

Regular (Perron) or *simple* (Olds) continued fractions [90][87] are ones in which all the numerators $a_i = 1$. They can be written even more concisely like this:

$$[b_0, b_1, b_2, \ldots b_n] \tag{4.68}$$

The elements a_i and b_i of a continued fraction are also known, respectively, as "partial numerator" and "partial denominator".

Every real number can be represented uniquely by a simple continued fraction. A number is rational if its continued fraction is finite and irrational if its continued fraction is infinite.

The second theorem immediately proves, for example, that all the numbers contained in the following list are irrational because their continued fractions are infinite. The proof of the irrationality of π which Johann Heinrich Lambert wrote in 1766 is also based on this theorem. Lambert showed that the continued fraction of $\arctan 1 =$

$\pi/4$ is infinite and therefore $\pi/4$ and hence π must be irrational (see page 192).

The transformation of a number from its decimal representation into its simple continued fraction form, and vice versa, is quite simple. It can be performed with any pocket calculator which has an inverse operation $1/x$. The two algorithms look like this:

```
// cf[] : array with the elements of the continued fraction
procedure NumberToCf(number, n, cf[0..n-1])
{
    for k:=0 to n-1
    {
        x       := floor(number)
        cf[k]   := x
        number:= 1/(number-x)
    }
}
```

and the reverse procedure for calculating a number from the elements of its simple continued fraction goes like this:

```
function CfToNumber(n, cf[0..n-1])
{
    number := cf[n-1]
    for k:=n-2 to 0 step -1
    {
        number := 1/number + cf[k]
    }
    return number
}
```

The simple continued fractions of some prominent constants are shown below[3]:

$$\phi = \frac{\sqrt{5}+1}{2} \quad \text{(Golden ratio)} \tag{4.69}$$
$$= 1.61803\,39887\,49894\,84820\ldots$$
$$= [1, 1, 1, 1, 1, 1, 1, 1, 1, 1, 1, 1, 1, 1, 1, 1, 1, 1, 1, 1, 1, \ldots]$$

$$\sqrt{2} = 1.41421\,35623\,73095\,04880\ldots \tag{4.70}$$
$$= [1, 2, 2, 2, 2, 2, 2, 2, 2, 2, 2, 2, 2, 2, 2, 2, 2, 2, 2, 2, 2, 2, \ldots]$$

$$\sqrt{3} = 1.73205\,08075\,68877\,29352\ldots \tag{4.71}$$
$$= [1, 1, 2, 1, 2, 1, 2, 1, 2, 1, 2, 1, 2, 1, 2, 1, 2, 1, 2, 1, \ldots]$$

---

[3] As this book is concerned almost exclusively with numbers that are transcendental and, as a very minimum, are irrational, we are naturally only interested in *infinite* continued fractions.

4.4 On continued fractions

$$e = 2.71828\,18284\,59045\,23536\ldots \tag{4.72}$$
$$= [2, 1, 2, 1, 1, 4, 1, 1, 6, 1, 1, 8, 1, 1, 10, 1, 1, 12, 1, 1, 14, 1, \ldots]$$

Euler, 1737

$$e^2 = 7.38905\,60989\,30650\,22723\ldots \tag{4.73}$$
$$= [7, 2, 1, 1, 3, 18, 5, 1, 1, 6, 30, 8, 1, 1, 9, 42, 11, 1, 1, 12, 54, \ldots]$$

Stieltjes, c. 1890 [90]

$$\sqrt{e} = 1.64872\,12707\,00128\,14684\ldots \tag{4.74}$$

Sundman, 1895

$$= [1, 1, 1, 1, 5, 1, 1, 9, 1, 1, 13, 1, 1, 17, 1, 1, 21, 1, 1, 25, 1, 1, \ldots]$$

$$\pi = 3.14159\,26535\,89793\,23846\ldots \tag{4.75}$$
$$= [3, 7, 15, 1, 292, 1, 1, 1, 2, 1, 3, 1, 14, 2, 1, 1, 2, 2, 2, 2, 1, 84, \ldots]$$

$$\sqrt{\pi} = 1.77245\,38509\,05516\,02729\ldots \tag{4.76}$$
$$= [1, 1, 3, 2, 1, 1, 6, 1, 28, 13, 1, 1, 2, 18, 1, 1, 1, 83, 1, 4, \ldots]$$

$$\pi^e = 22.45915\,77183\,61045\,47343\ldots \tag{4.77}$$
$$= [22, 2, 5, 1, 1, 1, 1, 1, 3, 2, 1, 1, 3, 9, 15, 25, 1, 1, 5, 4, 1, \ldots]$$

$$\sqrt[3]{2} = 1.25992\,10498\,94873\,16476\ldots \tag{4.78}$$
$$= [1, 3, 1, 5, 1, 1, 4, 1, 1, 8, 1, 14, 1, 10, 2, 1, 4, 12, 2, 3, 2, 1, \ldots]$$

$$\gamma = \lim_{n \to \infty} \sum_{k=1}^{n} \frac{1}{k} - \ln(n+1) \quad \text{(Euler's constant)} \tag{4.79}$$
$$= 0.57721\,56649\,01532\,86060\ldots$$
$$= [0, 1, 1, 2, 1, 2, 1, 4, 3, 13, 5, 1, 1, 8, 1, 2, 4, 1, 1, 40, 1, 11, 3, \ldots]$$

Euler, 1734

On closer examination we find there is a recurring pattern in the continued fractions of the first 6 constants, e.g. in $\sqrt{e}$ the pattern $1, 1, 4n+1$, whereas this does not occur in the second five cases beginning with π.

We cannot conclude from this that the simple continued fractions of many or even of the majority of common constants possess a regular pattern. In reality, this applies only to a minority, namely to the so-called quadratic irrationalities, one example of which is the "Golden ratio" ($\phi = 1 + \sqrt{5}/2$), also to a few well-known transcendental numbers, such as the constant e itself and individual algebraic expressions containing e, e^2 or $e^{1/q}$, where q is an integer.

Interestingly, in almost all irrational or transcendental numbers, including π, the geometric mean of their first n continued fraction elements converges towards a fixed limit and this limit is the same in all such numbers. This astonishing discovery was made in 1935 by Alexander Khintchine (1894–1959) [74]:

$$\lim_{n\to\infty} \sqrt[n]{b_1 \cdot b_2 \cdot \ldots \cdot b_n} = 2.68545\,20010\ldots = K_0 \qquad (4.80)$$

The limit K_0 is defined as $\prod_{k=1}^{\infty}[1 + \frac{1}{k(k+2)}]^{\log_2 k}$ and became known as the *Khintchine constant*. The number is difficult to calculate, and only a modest 110,000 decimal places are available.

For the number π, the geometric mean of the first $17,001,303$ elements of its simple continued fraction come to 2.686393 [12], which is fairly close to the Khintchine constant. On the other hand, the geometric mean for the number e, the big exception, moves away as n increases, while in the case of another exception, the constant ϕ (Golden ratio), whose continued fraction consists of pure ones, the geometric mean does converge, but towards 1 rather than K_0.

If one looks at the table with the approximations for π on page 54 in this light, one could form the opinion that continued fractions produce a more compact representation of π than the decimal representation. For the table shows, for example, that π can be just as well approximated with 3 continued fraction elements as with 6 decimal digits. However, this is only the case at particular places, and these are precisely the ones at which it is especially beneficial to terminate a continued fraction. Further on in both representations, the advantage of the continued fraction always disappears, and generally the simple continued fraction of an irrational number requires just as many elements as its equally accurate decimal fraction possesses decimal places. The quotient from the number of accurate decimal places of the approximation and the number of continued fraction elements converges towards a fixed limit which is close to 1, as Alexander Khintchine once again proved in 1935. The exact value $1.03064\ldots = \pi^2/(6 \cdot \ln 2 \cdot \ln 10)$ was stated somewhat later, in 1937, by Paul Lévy.

5. Arcus Tangens

5.1 John Machin's arctan formula

By 2000 BC the Babylonians and Egyptians had arrived at one decimal place approximations for π based on the *measurement* of circles. Archimedes was the first mathematician to develop a *geometry-based* approximation in 250 BC and his polygon method served as the basis for π approximations for the next 2000 years until (by 1630) 39 decimal places had been calculated.

The second half of the 17th century saw the development of infinitesimal analysis, which made it possible to formulate infinite expressions for π. This was an *analytical* method of calculation by means of which π could be investigated to a much greater depth. Among the infinite expressions for π, a sub-species of series which are based on the arctan function emerged.

The arcus functions are the inverse of the trigonometric functions (sin, cos, tan etc). As their name suggests, they constitute (circular) arcs or sections of them. If $x = \tan y$, then $\arctan x = y$, if y lies between $-\pi/2$ and $+\pi/2$. For our purposes the special case of $x = 1$ is particularly interesting as the arctan function for this case results in the value $\pi/4$ (or 45 degrees of arc).

$$\tan \frac{\pi}{4} = 1 \quad \text{hence,} \quad \frac{\pi}{4} = \arctan 1 \tag{5.1}$$

The arctan function possesses a series which is relatively easy to calculate and was originally discovered by James Gregory (1638?–1675). For the area under the curve $y = \frac{1}{1+x^2}$ in the interval $[0, x]$ he found:

$$\arctan x = \int_0^x \frac{dt}{1+t^2} \tag{5.2}$$

From this he derived the *Gregory Series* in 1671:

5. Arcus Tangens

$$\arctan x = x - \frac{x^3}{3} + \frac{x^5}{5} - \frac{x^7}{7} + \ldots \tag{5.3}$$

Gregory, 1671

To obtain a formula for π with this series (5.3), one needs only to set $x = 1$ because, as mentioned, $\arctan 1 = \frac{\pi}{4}$. The resulting series is called the *Leibniz series*:

$$\frac{\pi}{4} = 1 - \frac{1}{3} + \frac{1}{5} - \frac{1}{7} + \frac{1}{9} - \ldots = \sum_{n=0}^{\infty} (-1)^n \frac{1}{2n+1} \tag{5.4}$$

Leibniz, 1674

No doubt due to its simplicity, "every schoolboy" knows the Leibniz series. However, it is not suitable for numerical calculations of π because its terms get smaller only very slowly. If one terminates the series after the nth term, the absolute error, i.e. the difference between the series sum and the true value of π, is only $\approx 1/n$. Therefore, for example, even if one were to calculate 2 billion terms, one would only obtain 9 accurate decimal places of π.

The poor convergence of the Leibniz series (5.4) is thus due to the fact that it attempts to do too much in one go, so to speak. It determines $\pi/4$ by calculating a single arctan, i.e. a single arc. If instead one assembles $\pi/4$ in a suitable fashion from relatively short circular arcs, i.e. out of several arctan values, one arrives at formulae which enable π to be calculated significantly more quickly.

The simplest of such compound arctan formulae dates back to Leonhard Euler (1707–1783), and goes like this:

$$\frac{\pi}{4} = \arctan \frac{1}{2} + \arctan \frac{1}{3} \tag{5.5}$$

Euler, 1738

The formula can be proven using the known trigonometric identity

$$\tan(\alpha + \beta) = \frac{\tan \alpha + \tan \beta}{1 - \tan \alpha \tan \beta} \tag{5.6}$$

if one makes the substitutions $\tan \alpha = 1/2$, i.e. $\alpha = \arctan 1/2$ and $\tan \beta = 1/3$ i.e. $\beta = \arctan 1/3$.

The following diagram provides a geometric interpretation of this formula:

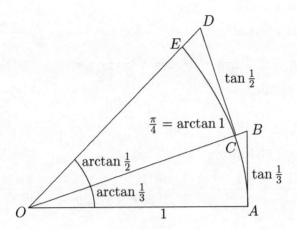

The sum of the arcs $AC = \arctan \frac{1}{3}$ and $CE = \arctan \frac{1}{2}$ produces the arc $AE = \arctan 1 = \frac{\pi}{4}$. If one replaces each of the arctan expressions in Euler's formula (5.5) with its Gregory series (5.3), one sees that *its* terms becomes smaller much more rapidly than those in the Leibniz series. For example the 100th term of the Leibniz series (16.59) (= arctan 1) yields just two leading zeros after the decimal point, but the 100th term of $\arctan \frac{1}{2}$ already produces 62, while the 100th term of $\arctan \frac{1}{3}$ produces as many as 98 zeros after the decimal point. This improvement in convergence offsets by far the disadvantage that two series have to be calculated now instead of one.

An even better formula can be obtained from two circular arcs as follows [76, p. 246]:

The number

$$\alpha = \arctan \frac{1}{5} = \frac{1}{5} - \frac{1}{3 \cdot 5^3} + \frac{1}{5 \cdot 5^5} - \frac{1}{7 \cdot 5^7} + \cdots \qquad (5.7)$$

is obtained from the Gregory series (5.3) if we set $x = \frac{1}{5}$. So, $\tan \alpha = \frac{1}{5}$. Therefore,

$$\tan 2\alpha = \frac{2 \tan \alpha}{1 - \tan^2 \alpha} = \frac{5}{12} \quad \text{and} \qquad (5.8)$$

$$\tan 4\alpha = \frac{120}{119} \qquad (5.9)$$

It will be seen that 4α is only a little greater than $\frac{\pi}{4}$. If we introduce a second angle β with $\beta = 4\alpha - \frac{\pi}{4}$, then

$$\tan\beta = \frac{\tan 4\alpha - \tan\frac{\pi}{4}}{1 + \tan 4\alpha \cdot \tan\frac{\pi}{4}} = \frac{1}{239} \tag{5.10}$$

β is very easy to calculate:

$$\beta = \arctan\frac{1}{239} = \frac{1}{239} - \frac{1}{3 \cdot 239^3} + \frac{1}{5 \cdot 239^5} - \cdots \tag{5.11}$$

α and β together produce

$$\frac{\pi}{4} = 4\alpha - \beta \tag{5.12}$$

$$= 4\arctan\frac{1}{5} - \arctan\frac{1}{239} \tag{5.13}$$

$$= 4\left[\frac{1}{5} - \frac{1}{3 \cdot 5^3} + \frac{1}{5 \cdot 5^5} - \cdots\right] - \left[\frac{1}{239} - \frac{1}{3 \cdot 239^3} + \cdots\right] \tag{5.14}$$

Machin, 1706

This is the so-called *Machin formula*. It bears the name of the person who discovered it, John Machin (1680–1752). He used it in 1706 to calculate π to 100 decimal places.

The successive terms of the series of the first summand in the Machin formula decline by around 1/25. The series therefore converges with $log_{10}25 = 1.39\ldots$ decimal places per series term. Due to the 5 in the denominator, it can also be calculated relatively easily using paper and pencil. The series of the second summand is more troublesome due to the "awkward" 239, but on the other hand it converges better, with around 4.76 places $(= \log_{10}(239^2))$ per term.

5.2 Other arctan formulae

Several systematic methods for deriving arctan relations exist. One of them is to use the formulas [41, p. 91]:

$$\arctan\frac{1}{a-b} = \arctan\frac{1}{a} + \arctan\left(\frac{b}{a^2 - ab + 1}\right) \quad \text{and} \tag{5.15}$$

Dodgson (Lewis Carroll) ?

$$\arctan\frac{1}{a} = 2\arctan\frac{1}{2a} - \arctan\frac{1}{4a^3 + 3a} \tag{5.16}$$

with appropriate choices for a and b. Substitution of $a = 2$ and $b = 1$ produces the above-mentioned arctan formula of Euler (5.5), while this formula plus substitution $a = 3$ and $b = 1$ results in the arctan formula (5.17) of Charles Hutton (1737–1823) shown below. Using this and other methods, over the years various further arctan formulae have been discovered for the result $\pi/4$, especially ones in which more than two arctan expressions occur. Here is a selection:

$$\frac{\pi}{4} = 2\arctan\frac{1}{2} + \arctan\frac{1}{7} \tag{5.17}$$

Hutton, 1776, Performance Index=3.28

$$= 3\arctan\frac{1}{4} + \arctan\frac{1}{20} + \arctan\frac{1}{1985} \tag{5.18}$$

Loney, 1893 [122], PI =2.73

$$= 6\arctan\frac{1}{8} + 2\arctan\frac{1}{57} + \arctan\frac{1}{239} \tag{5.19}$$

Størmer, 1896, PI=2.10

$$= 4\arctan\frac{1}{5} - \arctan\frac{1}{239} \tag{5.20}$$

Machin, 1706, PI=1.85

$$= 8\arctan\frac{1}{10} - \arctan\frac{1}{239} - 4\arctan\frac{1}{515} \tag{5.21}$$

Klingenstierna c. 1730, PI=1.79 [114, p. 296]

$$= 12\arctan\frac{1}{18} + 8\arctan\frac{1}{57} - 5\arctan\frac{1}{239} \tag{5.22}$$

Gauss [56, II, p. 524], PI=1.79

$$= 22\arctan\frac{1}{28} + 2\arctan\frac{1}{443} - 5\arctan\frac{1}{1393} -$$
$$- 10\arctan\frac{1}{11018} \tag{5.23}$$

Escott, PI=1.63

$$= 44\arctan\frac{1}{57} + 7\arctan\frac{1}{239} - 12\arctan\frac{1}{682} +$$
$$+ 24\arctan\frac{1}{12943} \tag{5.24}$$

Størmer, 1896, PI=1.59

This list contains some of the arctan formulae whose originators are known. There are many other such formulae, some of whose originators are known and others not. On our CD-ROM you will find a substantial collection of them in the `arith` directory, including some "monsters" with 11, 12 or even 13 terms, also an algorithm which can be used to automatically discover still more.

Along with the names of the originators, the list provides a "performance index" (PI). The smaller this is, the better, as this means that calculation of π requires less effort. The performance index is based on calculation of the expression $1/\log a_1 + 1/\log a_2 + \cdots + 1/\log a_n$.

From the list it is clear that the Machin formula is one of the best. It is also the best of the four possible [32, p. 345] arctan formulae which comprise only two terms. This explains why for over 250 years it was the favourite formula of π digit hunters.

Nevertheless, other arctan formulae were also widely used for a long time. For example, Leonhard Euler in 1755 used the following relation:

$$\frac{\pi}{4} = 5\arctan\frac{1}{7} + 2\arctan\frac{3}{79} \tag{5.25}$$

Euler, PI=1.89

Its performance index of 1.89 is worse than the formulae cited above. But Euler discovered an elegant way to reduce the the amount of work. He derived the identity

$$\arctan x = \frac{y}{x}\left(1 + \frac{2}{3}y + \frac{2\cdot 4}{3\cdot 5}y^2 + \frac{2\cdot 4\cdot 6}{3\cdot 5\cdot 7}y^3 + \cdots\right) \tag{5.26}$$

where $y = x^2/(1+x^2)$.

Euler, 1755

and inserted the arguments of (5.25) into it. The first one, $x = \frac{1}{7}$, leads to $y = \frac{2}{100}$, while the second one, $x = \frac{3}{79}$ produces $y = \frac{144}{10000}$. The powers of 10 which are now in these y's reduce their decimal computation to a matter of simple decimal shifts. By this means Euler was able to calculate 20 decimal places of π in less than an hour [32, p. 340].

In the previously mentioned arctan formula of Gauss (5.22), the arguments 1/18 and 1/57 have a similar useful property:

$$\arctan\frac{1}{18} = 18\left(\frac{1}{325} + \frac{2}{3\cdot 325^2} + \frac{2\cdot 4}{3\cdot 5\cdot 325^3} + \cdots\right) \tag{5.27}$$

and

$$\arctan\frac{1}{57} = 57\left(\frac{1}{3250} + \frac{2}{3\cdot 3250^2} + \frac{2\cdot 4}{3\cdot 3250^3} + \cdots\right) \tag{5.28}$$

Once again the work is simplified through decimal shifts, as a result of which the terms of the second series can be obtained from those of the first series. For this reason, the Gaussian arctan formula (5.22) was promoted to the "best formula for π calculations up to 1,000 decimal places" in the pre-computer era [16].

A π calculation using arctan formulae is very easy to programme. On our CD-ROM there is a quite elementary C program for doing this.

One of the authors (JA) devoted a considerable effort a few years ago to the search for arctan formulae, in the course of which he discovered some beauties (see formulae (16.124) to (16.132) in our collection of formulae beginning on page 233). He was looking for expressions in which the first term had the biggest possible denominator. He was especially pleased when his computer churned out the following arctan formula consisting of 11 terms:

$$\frac{\pi}{4} = 36462 \arctan \frac{1}{390112} + 135908 \arctan \frac{1}{485298} +$$
$$+ 274509 \arctan \frac{1}{683982} - 39581 \arctan \frac{1}{1984933} +$$
$$+ 178477 \arctan \frac{1}{2478328} - 114569 \arctan \frac{1}{3449051} -$$
$$- 146571 \arctan \frac{1}{18975991} + 61914 \arctan \frac{1}{22709274} -$$
$$- 6044 \arctan \frac{1}{24208144} - 89431 \arctan \frac{1}{201229582} -$$
$$- 43938 \arctan \frac{1}{2189376182} \qquad (5.29)$$

Arndt [7], 1993

The following two arctan formulae deserve a special mention for a quite different reason:

$$\frac{\pi}{4} = \arctan \frac{1}{2} + \arctan \frac{1}{5} + \arctan \frac{1}{13} + \arctan \frac{1}{34} + \cdots$$
$$= \sum_{n=1}^{\infty} \arctan \frac{1}{F_{2n+1}} \qquad (5.30)$$

and

$$\frac{\pi}{4} = \frac{3\sqrt{5} - 5}{2} - \sum_{n=1}^{\infty} F_{2n} \arctan \left(\frac{2}{3F_{2n+2} + F_{2n+2}^3} \right) \qquad (5.31)$$

Arndt, 1994

These series have an infinite number of arctan summands and are therefore no good for an effective π calculation. But their distinguishing feature is that they contain the famous *Fibonacci numbers* F_n. Fibonacci numbers are named after their discoverer, Fibonacci, whose real name was Leonardo of Pisa (1180–1240), and go $F_n = 1, 1, 2, 3, 5, 8, 13, 21, 34, \ldots$. Each succeeding number is the sum of its two predecessors $F_{n+2} = F_{n+1} + F_n, n \geq 1$.

Fibonacci numbers have various beautiful characteristics and occur often in nature. In particular, they forge a bridge between mathematics and art because the ratio of successive Fibonacci numbers converges towards the "Golden ratio" $\phi = \frac{1}{2}(\sqrt{5}+1) = 1.61803\ldots$. This Golden ratio ϕ has been regarded since antiquity as especially aesthetic in sculptures, paintings and buildings and logically occurs also in Arndt's formula (5.31). In this way Fibonacci numbers endow our π with an artistic touch.

6. Spigot Algorithms

The *spigot algorithm*, developed by Stanley Rabinowitz and Stanley Wagon [95], is a recent and elegant method of calculating π. The algorithm is ideal for running on a personal computer.

1. The spigot algorithm starts producing digits of π right from start-up and thereafter churns them out at regular intervals. Under all the other methods, π has to be calculated completely in a buffer and is only output at the end, all at once. By contrast, with the spigot algorithm the π digits trickle out one at a time. It is possible to watch it at work, so it is well suited for online demonstrations on the Internet, for example. The CD-ROM which goes with this book contains the Java applet `spigot/pispigot.htm`, which shows the algorithm in action. You only need to load it into a standard Java-capable browser.
2. The algorithm works with nice small integers; even for 15,000 digits of π the values of its variables do not climb to more than 32 bits (including plus and minus signs), so that the `long` C data type is sufficient on standard 16-and 32-bit compilers. This means there are no problems with rounding, deletion or truncation, which can make life miserable with other algorithms.
3. Implementation of the spigot algorithm does not require any extraneous software, such as, for example, a high precision library. Everything you need will be found in any standard C compiler.
4. The spigot algorithm is surprisingly fast. Although the time it requires is of quadratic order, so that it cannot compare with high-performance algorithms such as the Gauss AGM algorithm (see page 91), nevertheless it regularly outdoes algorithms that are based on arctan series (see page 69).
5. The mathematics behind the spigot algorithm is simple.

6. Spigot Algorithms

6. The spigot algorithm can be written in only a few lines of source code. The shortest π programs are based on it. A proof of this has already been provided above (see page 37).

6.1 The spigot algorithm in detail

The starting point is the following simple structured series for π:

$$\pi = 2 + \frac{1}{3}(2 + \frac{2}{5}(2 + \frac{3}{7}(2 + \cdots))) \qquad (6.1)$$

It can be derived from the Leibniz series (16.59) without great effort if one just uses the Euler transformation [76, p. 255]. However, we will not perform that step now.

The series (6.1) can be conceived of as a number in a number system with a variable base. Normally we encounter only numbers with a fixed base, which is usually base 10. Every place of such a number must be multiplied with a value which is higher than the value of the next digit to the right, by a constant factor known as the "base".

But occasionally we come across numbers in which this factor is not constant, e.g. the numbers which correspond to the expression "2 weeks, 3 days, 4 hours and 5 minutes". Because the ratio of weeks to days = 1 : 7, days to hours = 1 : 24 and hours to minutes = 1 : 60, when converting this number to a decimal number (with the unit "weeks"), three different factors have to be considered, namely 1/7, 1/24 and 1/60. Hence, to answer the question how many weeks are in the example, we have to calculate the following:

$$2 + \frac{1}{7}(3 + \frac{1}{24}(4 + \frac{1}{60}(5))) \qquad (6.2)$$

Now compare this expression with the right-hand side of the π series (6.1). You will see the same arrangement of brackets and also the different bases. On the other hand, in the π series all the numerals are equal, i.e. = 2, whereas in the week example they are different, i.e. = 2, 3, 4 and 5; the π series is also infinite, whereas the example terminates after the fourth digit.

Back to π. The task which the spigot algorithm performs consists simply of converting the π series (6.1) to our base 10 number system, i.e. to this form:

$$\pi = 3.1415\ldots = 3 + \frac{1}{10}(1 + \frac{1}{10}(4 + \frac{1}{10}(1 + \frac{1}{10}(5 + \cdots)))) \qquad (6.3)$$

6.1 The spigot algorithm in detail

Such a task is known in arithmetic as *radix conversion* and functions – in the present case – as follows:

On every step one π decimal place is calculated. To do this, first of all all the digits in the number to be converted are multiplied by 10 (the new base). Then, starting from the right, every decimal place is divided by the previous base $(2i+1)/i$ which applies to this decimal place. On every division, the remainder is retained and the integer quotient is carried over to the next decimal place. The most recently calculated carry-over is the new decimal place of π.

The question arises as to how many terms in the π series (6.1) really have to be carried over in order to obtain n decimal places of π, including the digit 3 before the decimal point. In their article, Rabinowitz and Wagon specify the value $\lfloor 10n/3 \rfloor$, where the notation $\lfloor x \rfloor$ in the customary manner refers to the largest integer $\leq x$, i.e. it means $\lfloor 10/3 \rfloor = 3$ and $\lfloor 3 \rfloor = 3$. They even "prove" that this value is "correct". Unfortunately, it is not correct, as they could have seen from the values $n = 1$ and $n = 32$. We are taking the liberty of correcting the mistake and take one more decimal place, i.e. $\lfloor 10n/3 + 1 \rfloor$ decimal places, in the (tested) supposition that we are then correct in every case.

Before we can begin to programme, we must explain the single real complication in the algorithm.

When the base of the π series (6.1) is changed, it is possible for a digit pair $= 10$ to occur. It can thus happen that at some point a 10 occurs at position p, i.e. $3.1415\ldots(p)(10)$. The 1 in the 10 is an unresolved carry-over and must be added to the previous digit i.e. $3.1415\ldots(p+1)0$. It can even happen that in front of such a 10 one or more 9s have been calculated, so that those places too must be corrected. The $3.1415\ldots(p)99\ldots9(10)$ must then be transformed into $3.1415\ldots(p+1)00\ldots00$.

This complication means that the π program cannot release the digits it has calculated immediately, but must store them in a buffer until the next digit or the next few digits has/have been calculated. When a new digit arrives, one or more digits will still be held in the buffer. The first of these will definitely be < 9 and the others, if there are any, exactly $= 9$. In this way we have the following situation in the buffer at the point when the new digit q arrives:

$$p\underbrace{99\ldots9}\leftarrow q$$

There are now three possibilities:

1. $q < 9$: it turns out not to have been necessary to retain the value in the buffer. p and any succeeding $99\ldots 9$ can be output as they are. q becomes the next decimal digit of p.
2. $q = 9$: no decision has been made so the number of 9s held in the buffer is increased by 1 as q is added to it.
3. $q = 10$: it turns out to have been necessary to store the intermediate values in the buffer because a 1 must now be added to the sequence of digits held in the buffer. As a result, p is increased by 1, and all the 9s held temporarily become 0s. All these decimal places are now ready and can be output, but the 1 in the 10 is stored temporarily, i.e. p is set to 0.

The complication thus arises because the π series (6.1) is not unique in relation to the second positions. For example, $\frac{2}{3}$ can be represented in two ways, namely by $0+\frac{1}{3}(0+\frac{2}{5}(2+\frac{3}{7}(3+\cdots)))$ and by $0+\frac{1}{3}(2)$. There are series for π which are unique, but they are much more long-winded to calculate than (6.1).

6.2 Sequence of operations

We have now assembled all the elements required to describe the sequence of operations in the spigot algorithm.

The spigot algorithm calculates the first n decimal places of π. It works with a field $a[0], a[1], \ldots, a[N]$ consisting of $N+1 = \lfloor (10n)/3 \rfloor + 1$ integers, where $\lfloor x \rfloor$ signifies the largest integer $\leq x$. Moreover, two variables p and q are used to record the first and the current provisional decimal place, also a numerator *nines* for the number of temporary nines.

Initialisation: set $p = 0$ and *nines* $= 0$.
 For $i = 0, 1, 2, \ldots, N$, set $a[i] = 2$.
Iteration: repeat until n decimal places have been output.
- *Multiply with the new base.* Multiply each $a[i]$ by 10.
- *Normalise.* Beginning on the right, from $i = N$ to $i = 1$, divide $a[i]$ by $(2i+1)$ to obtain a quotient q and a remainder r. Replace $a[i]$ with r. Multiply q by i and add the result (the carry-over) to the element $a[i-1]$.

- *Calculation of the next provisional digit of π.* The left-hand digit $a[0]$ is post-processed. It is divided by 10. The division remainder replaces $a[0]$, while the quotient q produces the next provisional digit of π.
- *Correct the old provisional digits.* If q is neither 9 nor 10, then the first provisional digit up to now, p, and the *nines* which succeed it are confirmed and output. The new first provisional digit p now becomes $= q$, and *nines* is set to $= 0$.

  If $q = 9$, then out of the provisional nines only the number *nines* is incremented by 1; no digits are output.

  If $q = 10$, the first provisional digit up to now, p, is increased by 1 and output. The provisional *nines* become zeros; they are likewise output. The new first provisional digit, p, becomes $= 0$ (that is the 1st digit of q). *nines* is reset to 0.

A small improvement can be achieved if one initialises the first provisional digit of p with a negative value and then intercepts this value during output. This means that the output begins immediately with 314... instead of initially with a zero.

The following C function spigot() is described in the sequence of operations below.

```
/*
 * function
 *        void spigot(digits)
 * Spigot program for pi
 * 1 digit per loop
 */

#include <stdio.h>
#include <stdlib.h>

void spigot(int digits)
{
    int    i, nines = 0;
    int    q,                        /* next prelim. digit                  */
           p = -1;                   /* previous prelim. digit              */
    int    len = 10*digits+3+1;      /* len: One more than R+W              */
    int    *a;                       /* array pointer                       */

    a = malloc(len*sizeof(*a));
    for (i=0; i < len; ++i)          /* Init a[] with 2's                   */
        a[i] = 2;
    while (digits >= 0)
    {                                /* Compensate for the very first digit */
        q = 0;
        for(i=len; --i >= 1; )
        {
            q += 10L * a[i];         /* q = carry + 10*a[i]                 */
            a[i] = q % (i+i+1);      /* a[i] := q % (2i+1)                  */
```

```
                q /= (i+i+1);           /* carry := floor(q,2i+1)*i          */
                q *= i;
            }
                                        /* first digit                       */
            q += 10L * a[0];            /* q := carry + 10 * a[0]            */
            a[0] = q % 10;              /* a[0] = q mod 10                   */
            q /= 10;                    /* q : next prelim digit             */
            if (q == 9)
                ++nines;                /* q == 9: increment no of 9's       */
            else
            {                           /* q != 9: print prelim. digits      */
                if (p >= 0)
                    printf("%01ld", p + q/10);   /* p : prev. prel. digit */
                if (digits < nines)     /* adjust digits to print            */
                    nines = digits;
                digits -= (nines+1);
                while (--nines >= 0)    /* print 9's or 0's                  */
                    printf(q == 10? "0" : "9");
                nines = 0;
                p = (q == 10 ? 0 : q);  /* set previous prelim. digit        */
            }
        }
    }
    free(a);
    return;
}
```

At the end of the much-cited article [95] by Rabinowitz and Wagon there is a PASCAL program which evidently was not written by the authors themselves but by a student, to whom they express their thanks. This program has also been typed out and tested by some readers of this book. They had various problems with it, for example, on 16-bit PASCAL compilers the program stops working from $n > 262$ because an integer overflow occurs at that point, or else at $n = 1$ and $n = 32$ it prints an incorrect last digit because the series length is too short, or else in many cases of n it outputs fewer than n digits because variable **nines** at the end of the program is not yet 0. Our above program attempts to avoid these weaknesses.

6.3 A faster variant

Two improvements can be made to the spigot algorithm which make it considerably shorter and quicker to run.

First of all, instead of π, 1000π is calculated, and hence the series

$$1000\pi = 2000 + \frac{1}{3}\left(2000 + \frac{2}{5}\left(2000 + \frac{3}{7}(2000 + \cdots)\right)\right) \qquad (6.4)$$

is used. The conversion is thus not performed in base 10 but in base 10,000, so that on every pass the program produces four decimal places instead of only one.

This trick not only has the effect that the entire program becomes 4 times faster, but an even more important effect is that the "complication" mentioned above becomes a lot simpler. The faster variant always waits only exactly 1 position (consisting of 4 digits) instead of having a variable number of places. This is sufficient for the first approx. 50,000 digits of π, as up to that point there is no case in which more than one such 4-digit chain has to be placed in a buffer. This is only necessary if 4 zeros occur one after the other in a position that is divisible by 4, and the first time that this occurs in π is at digit position 54,936.

The second improvement is really a textbook tip which, however, was evidently forgotten in the original formulation of the spigot algorithm. Namely, with a radix conversion, after each result place, one can shorten the remainder that still has to be converted, by the number of bits of the result place. Hence after each calculation of one place consisting of 4 decimal digits, the length of the f[] field can be reduced by $\lfloor 10\cdot 4/3+1 \rfloor = 14$ places. This improvement speeds up the program by an additional factor of 2.

Here is a C program for the faster variant. It is an expanded version of the mini-program shown on page 37.

```
/*
 * Spigot program for pi to NDIGITS decimals
 * 4 digits per loop
 * Expanded version
 * Thanks to Dik T. Winter and Achim Flammenkamp.
 */

#include <stdio.h>
#include <stdlib.h>

#define NDIGITS 15000           /* max. digits to compute  */
#define LEN     (NDIGITS/4+1)*14 /* nec. array length      */

long a[LEN];                    /* array of 4 digit-decimals*/
long b;                         /* nominator prev. base    */
long c = LEN;                   /* index                   */
long d;                         /* accumulator and carry   */
long e = 0;                     /* save prev. 4 digits     */
long f = 10000;                 /* new base, 4 dec. digits */
long g;                         /* denom prev. base        */
long h = 0;                     /* init switch             */

int main(void)
{
    for ( ; (b=c-=14) > 0; )    /* outer loop:4 digits/loop*/
    {
        for (; --b > 0; )       /* inner loop: radix conv  */
        {
```

```
            d *= b;                  /* acc *= nom. prev base    */
            if (h == 0)
                d += 2000 * f;       /* first outer loop         */
            else
                d += a[b] * f;       /* non-first outer loop     */
            g=b+b-1;                 /* denom prev. base         */
            a[b] = d % g;
            d /= g;                  /* save carry               */
        }
        h = printf("%04ld", e+d/f);  /* print prev 4 digits */
        d = e = d % f;               /* save current 4 digits    */
                                     /* assure a small enough d */
    }
    return 0;
}
```

On the basis of the NDIGITS definition, this program calculates exactly 15,000 decimal places of π. There is no reason why it has to stop at this limit if one is not concerned about the portability of the program and compliance with ANSI C. As most C compilers around ignore an *overflow* in the evaluation of integer expressions which depend on plus and minus signs and there is sufficient "breathing space" in the array length, this program can normally be expanded to calculate twice as many π positions (using #define NDIGITS 32500). This number was still a world record 40 years ago. Even more decimal places can be calculated with floating point variables and arithmetic.

But however large a data range you select, even with this variant of the spigot algorithm, i.e. where each place consists of only 4 π digits and only one such place is held in the buffer, it is still not possible to get past π digit position 54932. This can only be achieved by removing at least one of these limiting conditions; the easiest one to remove is the first.

6.4 Spigot algorithm for e

The spigot algorithm is clearly not limited to the calculation of π.

Let us consider the transcendental number $e = 2.7182\ldots$. Its series expansion which is the equivalent of the above π series (6.1) goes as follows:

$$e = 1 + \frac{1}{1}(1 + \frac{1}{2}(1 + \frac{1}{3}(1 + \cdots))) \tag{6.5}$$

Here, as in the π series, different bases occur, but this time they are all of the type that their numerator = 1. This means that when calculating the carry-over, there is no need for a multiplication operation.

More important, however, is the fact that the e series (6.5) is unique, so that the complication which occurs with π does not occur here. A spigot program for the decimal places of e is therefore simpler.

Here is a 138-character long program for the calculation of e to 15,000 places in the style of the mini-π program shown on page 37:

```
/* note: N=15000, LEN=87700 >= 1.4*N*log10(N), 84700=LEN-N/5 */
a[87700],b,c=87700,d,e=1e4,f=1e5,h;
main(){for(;b=c--,b>84700;h=printf("%05d",e+d/f),e=d%=f)
for(;--b;d+=f*(h?a[b]:e),a[b]=d%b,d/=b);}
```

Back on page 36 we promised to tell you how Lievaart's "obfuscated" program works. Well, it too works with the spigot algorithm for e.

7. Gauss and π

One of the fastest method of calculating π, if not the fastest of all, is almost 200 years old. It was invented by the German mathematician Carl Friedrich Gauss (1777–1855) around 1800. It was subsequently forgotten and only unearthed 170 years later, when two researchers, Eugene Salamin [100] and Richard Brent [37] independently rediscovered it at the same time and turned it into the basis for superfast π calculations.

Gauss's calculation method has since formed the basis of a number of other formulations, as one can tell from the different names by which the method is known in the literature. Thus we find the "Brent-Salamin iteration" or the "Gauss-Legendre method" and other variations. In this book, we refer to the method as the *Gauss AGM algorithm* as it bears Gauss's unmistakable hallmark and its distinctive feature is the arithmetic geometric mean (AGM).

7.1 The π AGM formula

Gauss's formula that was so important in the study of π is shown in its original form on page 101. Using modern notation, it looks like this:

$$\pi = \frac{2\,\mathrm{AGM}^2(1, \frac{1}{\sqrt{2}})}{\frac{1}{2} - \sum_{k=1}^{\infty} 2^k c_k^2} \tag{7.1}$$

Gauss, 1809, Salamin, Brent, 1976

The essential element in this formula is the function $AGM(a,b)$, which produces the so-called *arithmetic-geometric mean* (AGM) of two numbers. The AGM is a combination of the arithmetic and the geometric means. The second variable, the c_k in the sum in the denominator, is directly linked with the AGM.

In everyday life we come across several 'means'. The arithmetic mean $(a+b)/2$ of two numbers a and b is a common tool and can be used, for example, to arrive at the average of two marks in a certificate of academic achievement. The geometric mean $\sqrt{a \cdot b}$ is also well-known and is used to calculate the average of two or more variables which are linked by a multiplication operation, for example to calculate the average of two interest rates. In both cases, the word "mean" refers to a variable which lies "in the middle" between two starting values.

The AGM is also the average of two variables a and b. However, this average cannot be calculated in one step, but requires an infinite number of steps, each of which approximates it more closely. The AGM calculation rule is "iterative", in that each step builds on and improves on the results of the previous step.

Algorithm (7.2) AGM rule (alg 7.2)

Initialise:
$$a_0 := a$$
$$b_0 := b$$
Iterate: $(k = 0, 1, 2, \ldots)$
$$a_{k+1} := \frac{a_k + b_k}{2} \xrightarrow{2} \mathrm{AGM}(a,b)$$
$$b_{k+1} := \sqrt{a_k \cdot b_k} \xrightarrow{2} \mathrm{AGM}(a,b)$$

Then converge a_k and b_k to the same limit $\mathrm{AGM}(a,b)$.

It can be seen that the a_{k+1} is calculated by taking the arithmetic mean of its predecessors a_k and b_k, and the b_{k+1} by taking their geometric mean.

When Gauss first got interested in the AGM (at the age of 14) he ascertained its numeric behaviour in a number of cases "manually". One of four surviving numeric examples begins with the starting values $a = \sqrt{2}$ and $b = 1$ and, applying the rule (alg 7.2), it proceeds as follows: [56, III, p. 364][1].

Arithmetic mean	Geometric mean	Accurate places
$a\ \ = 1.41421\,35623\,73095\,04880\,2$	$b\ \ = 1.00000\,00000\,00000\,00000\,0$	0
$a_1 = 1.20710\,67811\,86547\,52440\,1$	$b_1 = 1.18920\,71150\,02721\,06671\,7$	0
$a_2 = 1.19815\,69480\,94634\,29555\,9$	$b_2 = 1.19812\,35214\,93120\,12260\,7$	4
$a_3 = 1.19814\,02347\,93877\,20908\,3$	$b_3 = 1.19814\,02346\,77307\,20579\,8$	9
$a_4 = 1.19814\,02347\,35592\,20744\,1$	$b_4 = 1.19814\,02347\,35592\,20743\,9$	19

---

[1] Gauss demonstrated from such columns of numbers that, mathematical genius through he was, he too did not shy away from laborious numeric calculations. Mathematicians today still stand in awe of him for this reason.

In the consecutive iteration steps, a_k and b_k move towards each other, a_k getting smaller and b_k getting bigger. (They would still move towards each other after crossing even if a had been smaller than b at the beginning). The value to which they converge, i.e. their limit values, is called the "AGM of a and b" and is written as AGM(a,b). In the case of $a = \sqrt{2}$ and $b = 1$ the limit value is clearly $1.19814\ldots$.

The AGM always lies between the geometric mean and the arithmetic mean of the starting values.

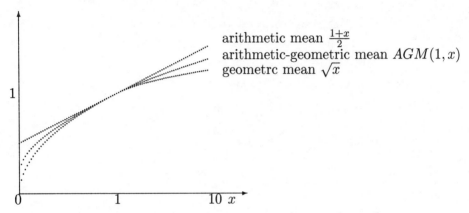

The main point to emerge from this numeric example of Gauss's is that a_k and b_k rapidly converge towards each other. The last column shows the number of decimal places which are identical for both variables. After only four iterations AGM$(\sqrt{2}, 1)$ is already accurate to 19 decimal places.

In each iteration, the number of correct decimal places approximately doubles. This *quadratic convergence* is the outstanding quality of the AGM and is what makes it so interesting to mathematicians. In our π formula (7.1) the speed at which π is calculated is determined by the AGM function (in the numerator) effectively "bequeathing" this quadratic convergence to it.

The denominator of formula (7.1) contains the variable c_k. This is a useful auxiliary variable which crops up again and again in the context of the AGM. The variables a_k and b_k are generated during the AGM iterations. The auxiliary variable c_k is defined as half the difference between them.

$$c_{k+1} := \frac{1}{2}(a_k - b_k) \tag{7.3}$$

identical with

$$c_{k+1}^2 = a_{k+1}^2 - b_{k+1}^2 = (a_{k+1} - a_k)^2 \tag{7.4}$$

As a and b move towards each other, c obviously converges towards 0.

Some of the basic features of the AGM are the following: With real arguments a_0 and b_0, $0 < b_0 \leq a_0$ and the auxiliary variable $c_{k+1} := (a_k - b_k)/2$, the following apply [32, p. 1–4]:

$$a_k = a_{k+1} + c_{k+1}$$
$$b_k = a_{k+1} - c_{k+1}$$
$$c_k = (a_k - b_k)/2 = c_{k-1}^2/(4a_k)$$
$$c_k^2 = a_k^2 - b_k^2 = (a_k - a_{k-1})^2$$
$$0 \leq a_{k+1} - b_{k+1} = (a_k - b_k)^2/(2(\sqrt{a_k} + \sqrt{b_k})^2)$$
$$M(a,b) := \lim_{k \to \infty} a_k = \lim_{k \to \infty} b_k$$
$$\text{AGM}(\lambda \cdot a, \lambda \cdot b) = \lambda \cdot \text{AGM}(a,b)$$
$$\text{AGM}(a,b) = \text{AGM}(\frac{a+b}{2}, \sqrt{ab})$$
$$\text{AGM}(1,b) = \frac{1+b}{2} \text{AGM}(1, \frac{2\sqrt{b}}{1+b})$$

Also:

$$b_k \leq b_{k+1} \leq a_{k+1} \leq a_k$$

and:

$$\text{AGM}(a_k, b_k) = \text{AGM}(a_{k+1}, b_{k+1}) = \text{AGM}(a,b)$$

The AGM is a well-defined quadratically convergent iteration for starting values $a_0 := 1, b_0 := z$, where $re(z) > 0$.

The AGM is also well defined for negative indices k: $a_{-k} = 2^k a_k^*$, $b_{-k} = 2^k c_k^*$ and $c_{-k} = 2^k b_k^*$, where a_k^*, b_k^* and c_k^* are generated from the AGM algorithm (alg 7.2) commencing with $a_0^* := a_0$, $b_0^* := c_0$ and $c_0^* := b_0$; note the "exchange" of b and c in the last two expressions.

7.2 The Gauss AGM algorithm

From his research into the AGM and various other mathematical deliberations which we will go into below, Gauss arrived at his formula for π which is so important today (7.1). It can be converted directly into an algorithm for π.

Algorithm (7.5) Gauss AGM (alg 7.5)

Initialise:
$$a_0 := 1$$
$$b_0 := 1/\sqrt{2}$$
$$s_0 := 1/2$$

Iterate ($k = 0, 1, 2, \ldots, K-1$)
$$t := a_k$$
$$a_{k+1} := (a_k + b_k)/2$$
$$b_{k+1} := \sqrt{tb_k}$$
$$c_{k+1}^2 := (a_{k+1} - t)^2$$
$$s_{k+1} := s_k - 2^{k+1} c_{k+1}^2$$

Compute an approximation of π:
$$\pi_K = \frac{(a_K + b_K)^2}{2 s_K}$$

The algorithm begins with initialisation of three variables a, b and s. In succeeding iterations the variables a and b are expanded further according to the AGM rule (alg 7.2) into an approximation for the term $\mathrm{AGM}(1, 1/\sqrt{2})$ which occurs in the denominator of (7.1). At the same time the sum s in the denominator is calculated using the above-mentioned auxiliary variable $c_{k+1}^2 = (a_{k+1} - a_k)^2$. The temporary variable $t = a_k$ is necessary because the value a_k is used twice more after it has already been replaced by a_{k+1}. The iteration runs either until a predefined level of precision has been achieved in the results or else, as here, until a specified number K of iterations has been performed. At the end of the procedure, the approximation for π is calculated in a single step outside the loop. Here, instead of the (unachievable) value $\mathrm{AGM}(1, 1/\sqrt{2})$, the next value $a_{K+1} = (a_K + b_K)/2$ is used. This is the most accurate approximation which can be achieved in K iterations.

These few lines thus define one of the best algorithms for calculating π. The first three iterations alone produce 19 accurate decimal places:

Iteration	p_K	Accurate places
1	3.14...	3
2	3.14159 26...	8
3	3.14159 26535 89793 238...	19
4	ditto	41
5	ditto	84
6	ditto	171
7	ditto	345
8	ditto	694
9	ditto	1392

Every extra iteration produces more than twice as many places as its predecessor. With 10 iterations the number of accurate decimal places increases 1000-fold and only 36 iterations are required to calculate π to 206.1 billion decimal places. The absolute error is [32, p. 48]:

$$\pi - p_K \leq \frac{\pi^2 2^{K+4} e^{-\pi 2^{K+1}}}{\text{AGM}^2(1, 1/\sqrt{2})} \text{ or}$$

$$\pi - p_{K+1} \leq \frac{(\pi - p_K)^2}{2^{K+1} \pi^2} \tag{7.6}$$

The quadratic decrease in the absolute error $(\pi - p_K)$ will be evident from the second expression, and from this one can determine that p_K is always smaller than π.

7.3 Schönhage variant

The calculation operations with long variables within the iteration loop are a decisive factor in determining the performance of the Gauss AGM algorithm, especially the long multiplication $a_k b_k$, the calculation of the square root of this, and also the squaring operation $(a_{k+1} - a_k)^2$. (On the other hand the long additions are of less consequence, likewise the multiplications by powers of 2, as these only involve decimal shifts of the variables which in any case are represented as binary numbers.)

Any improvement in performance is of course highly desirable. The following version, devised by Arnold Schönhage [105, p. 266], significantly speeds up the Gauss AGM algorithm. This variation avoids the long multiplication $a_k b_k$ required on every iteration.

Algorithm (7.7) Gauss AGM, Schönhage variant (alg 7.7)

Initialise:
$$a_0 := 1$$
$$A_0 := 1$$
$$B_0 := 0.5$$
$$s_0 := 0.5$$
Iterate: $(k = 0, 1, 2, \ldots, K - 1)$
$$t := (A_k + B_k)/4$$
$$b_k := \sqrt{B_k}$$
$$a_{k+1} := (a_k + b_k)/2$$
$$A_{k+1} := a_{k+1}^2$$
$$B_{k+1} := 2(A_{k+1} - t)$$
$$s_{k+1} := s_k + 2^{k+1}(B_{k+1} - A_{k+1})$$
Compute the approximation of π:
$$p_K = \frac{A_K + B_K}{s_K}$$

Schönhage dispenses with the long multiplication of $a_k b_k$ by replacing that product with less time-consuming operations involving other variables which occur: $a_k b_k = 2(a_{k+1}^2 - \frac{1}{4}(a_k^2 + b_k^2))$. (The squares of a and b are stored in the variables A and B.) The resulting performance is around 25% better. This is achieved without any implementation tricks simply through rearrangement of the operations.

In this way Schönhage avoids another long square root calculation during initialisation and, during the calculation of the approximation for π, he avoids a long squaring operation.

If one takes advantage of the fact that the b_k and B_k can share the same memory store, then the Schönhage arrangement, like the original form, requires only 5 long variables, namely a, A, $b = B$, t and s, thus avoiding any storage penalty for the speed gained.

It is very easy to convert the Gauss AGM algorithm to a program, using the Schönhage variation:

```
a = A = 1
B = s = 0.5    // B and b can share storage
for k=1 to N
    t = (A+B)/4
    b = sqrt(B)
    a = (a+b)/2
    A = a^2
    B = (A-t)*2
    s = s + (B-A) * 2^k
end for
pi = (a+b)^2/(2*s)
```

For the first few digits of π this program can be implemented without any problems in a standard programming language with normal floating point arithmetic or, better still, with a computer algebra system. Calculation of world record numbers of decimal digits is not, however, so easy and major implementation problems arise.

Mass storage problem. The long variables a, A, b=B, s and t have to be sized from the beginning to match the length of the result; but when one is dealing with hundreds of billions of π digits, the main memory of even the biggest computers is not sufficient, so that the variables have to be stored outside of main memory and introduced into it in stages. To do this quickly, considerable processor-specific work must be done on memory management.

High precision arithmetic. Computers possess only a (very) limited machine precision in their instructions, and many can only handle variables a maximum of 16 (decimal) places long. Hence "long" operations such as addition or modification of long variables have to be assembled from "short" operations. The procedure is theoretically complex and its practical implementation is performance-critical. For this reason the "high precision arithmetic" that is required is generally the most difficult and the most challenging part of a π program. It usually constitutes the very reason for developing a program in the first place. However, it is also possible to "purchase" this basic work (including the memory management functions mentioned above) and draw it from one of the many high precision libraries (such as our hfloatlibrary), by integrating such a library into one's program in an appropriate fashion.

7.4 History of a formula

The AGM

The Gaussian π formula (7.1) looks relatively harmless but there is actually more to it than meets the eye. It is remarkable not just because it is so effective but also for its mathematical background and the history of how it came about.

The arithmetic-geometric mean (AGM) and its calculation rule are less old than one might suppose from its simplicity. Both were discovered some 200 years ago. Surprisingly, the calculation rule for the

AGM is actually older than the AGM itself [46]. The French mathematician Joseph-Louis Lagrange (1736–1813) was the first to use the rule (alg 7.2) in 1785 for the calculation of approximations for elliptic integrals, although he did not discover the AGM or its relationship with elliptic integrals himself. It was Gauss who succeeded in making these discoveries. As David Cox writes, "... we have the amusing situation of Gauss, who anticipated so much in Abel, Jacobi and others, himself anticipated by Lagrange." [45, p. 315].

It was in 1791, at the tender age of 14, that Gauss discovered the AGM. His discovery exercised such a fascination on him that during the next 10 years he worked almost without interruption on developing an AGM theory and took this work to heights which have not been surpassed since [57, p. 186].

Then when he was 22 or 23, Gauss wrote an essay (in Latin) on his discoveries relating to the AGM: *de origine proprietatibusque generalibus numerorum arithmetico-geometricorum* (On the origin and general characteristics of arithmetic-geometric means) [56, III,pp. 361–374]. However, he never had the essay published and it was only printed in 1866 as part of his estate. Gauss's lifetime publications included only one piece on the AGM, and that was in 1818 in his treatise *Determinatio attractionis* (On the attraction of the elliptic ring) [56, III, p. 352–353], which contains the third proof of the fundamental identity (7.13). Additional information is provided in his formal writings.

Very little is known about Gauss's teenage research on the AGM. However, it seems likely that he possessed the basic knowledge quoted above very early, and evidently he knew from his research on the lemniscate that special significance is attached to the argument pair $\sqrt{2}$ and 1. There are also signs that already in 1794 he was aware of the connection between the AGM and "numerical integrals", as he called them, which today are referred to as theta functions[2].

There were two phases to Gauss's research on the AGM, 13 May 1799 marking the start of the second phase. It was on this that Gauss, now 22 years old, succeeded in forging a link between the AGM and a quite different area which up to then he had been following quite separately.

---

[2] Theta functions are power series whose exponents are square roots.

The lemniscate

The second constituent in Gauss's π formula (7.1) besides the AGM is the so-called lemniscate function. This function take its name from the *lemniscate*[3], a curve which Gauss had already studied in detail as a teenager. The lemniscate takes its name from the Greek word lemniskos, meaning a little ribbon or bow. It looks rather like a sideways 8:

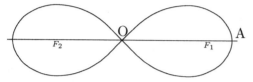

Expressed in polar co-ordinates, its equation is as follows:

$$r^2 = a^2 \cos 2\theta \qquad (7.8)$$

When Gauss turned to the lemniscate, this curve was almost exactly 100 years old. Two brothers had discovered it independently in the same year, 1694, Jakob Bernoulli (1654–1705) and his younger brother Johann (1667–1748). Jakob's publication about it appeared in September 1694 and Johann's only a month later. Following publication, the brothers had a violent dispute about which of them could claim to have invented it [45, p. 311]. The quarrel went so far that Johann swore he would never return to Basle as long as his brother was still living[4].

Jakob Bernoulli had discovered the lemniscate via the so-called *elastic curve* , and still expressed it in Cartesian co-ordinates ($xx + yy = a\sqrt{xx - yy}$). This elastic curve is formed if one forces a rod to bend so far that its ends are perpendicular to an imaginary line connecting their endpoints.

---

[3] The lemniscate is the set of all the points for which the product of the distances of two fixed points F_1 and F_2 possesses the value $(\overline{F_1 F_2}/2)^2$.

[4] When Jakob died in 1705, Johann succeeded him to his professorship in Basle, which post he held for another 43 years [83, p. 112].

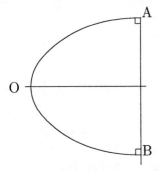

The equation for the elastic curve is

$$y = \int_0^x \frac{z^2\, dz}{\sqrt{1-z^4}} \tag{7.9}$$

More important historically than the equation for the curve, however, was the equation for its arc length AOB. This goes like this:

$$\varpi = 2\int_0^1 \frac{dz}{\sqrt{1-z^4}} = 2.6220575542\ldots \tag{7.10}$$

Jakob Bernoulli succeeded in writing a formula for this arc length in 1691. He needed it again three years later when he was looking for a curve that was analytically more manageable but had to have the same arc length. He found this curve in 1694 in the lemniscate, whose half perimeter is in fact given by the integral (7.10). Thus one can say that the arc length of the lemniscate was known before the lemniscate itself [45, p. 311].

In the eighteenth century, even before Gauss, the elastic curve and the lemniscate figured in many mathematical treatises. For example, in 1730 James Stirling (1692–1770) specified approximate values for the integrals (7.9) on the interval $[0,1]$ and for (7.10), which were accurate to 17 places. Giovanni Fagnano (1715–1797) discovered some methods for dividing up the lemniscate arc into n equal parts, where n can be equal to $2^m, 3 \cdot 2^m$ or $5 \cdot 2^m$ [45, p. 313].

From 1748 Leonhard Euler (1707–1783) developed a first stage of the theory of elliptic integrals, starting from the lemniscate. In particular, Euler discovered the remarkable relationship between the two integrals already cited:

$$\int_0^1 \frac{dz}{\sqrt{1-z^4}} \cdot \int_0^1 \frac{z^2\, dz}{\sqrt{1-z^4}} = \frac{\pi}{4} \tag{7.11}$$

Gauss began work on the lemniscate in January 1797 (at the age of 19), as is borne out by entry no. 51 in his mathematical diary [55, p. 67]: "Curvam lemniscatam a

$$\int \frac{dx}{\sqrt{1-x^4}}$$

pendentem perscrutari coepi." (I have begun to research the lemniscate curve, which is a function of ...). Originally the word "elasticam" had occurred at this point, but Gauss later crossed it out and replaced it with "lemniscatam". Soon afterwards he began using the symbol ϖ, as used in (7.10).

Gauss began his research with the problem of how to divide a lemniscate into equal parts which was mentioned above in connection with Fagnano. This took him to the "lemniscate functions" sinlemn($\int_0^x (1-z^4)^{-1/2}\,dz$) = x and coslemn($\varpi/2 - \int_0^x (1-z^4)^{-1/2}\,dz$) = x, which he defined for complex numbers and among which he found several relationships. He also found (one year after discovering the regular 17-sided polygon) a construction using only compass and straight edge, by means of which the lemniscate can be divided into 5 (not 17) equal parts[5]. The first three months of 1797 saw a large number of discoveries in this area, as indicated by four diary entries (nos. 54, 59, 60, 62).

Gauss was particularly encouraged when he noticed the analogy of the lemniscate functions to the circle functions, for example the following:

$$\varpi/2 = \int_0^1 (1-z^4)^{-1/2}\,dz \text{ and } \pi/2 = \int_0^1 (1-z^2)^{-1/2}\,dz \qquad (7.12)$$

The analogy is visually even greater when one reflects that in Gauss's day the symbol Π was often still used in lieu of the symbol ϖ.

In 1798 Gauss began to study the expression $\frac{\pi}{\varpi}$, which turned out to be the key to further discoveries. Amongst other things he found a series for the inverse of this expression, and with it he calculated ϖ/π to 15 decimal places.

Suffice to say that by July 1798 Gauss already knew "everything" about the quotient π/ϖ [45, p. 319]. Entry no. 92 of his diary states,

---

[5] Henrik Abel (1802–1829) subsequently continued Gauss's work and found that exactly the same ratios apply to the lemniscate as to the circle and, in particular, that division of the lemniscate into 17 parts is possible with compass and straight edge too.

"We have gained some very elegant details about the lemniscate, which have exceeded all expectations, and indeed using methods which open up an entirely new field." He sensed that he had made an important discovery and under no. 95 (in October 1798) he even wrote, "A new field of analysis stands before us, namely the investigation of functions etc." Gauss was so excited that he left his sentence unfinished.

Merger of the two fields

Then came 30 May 1799. Gauss noted: "That the AGM is equal to $\frac{\pi}{\varpi}$ between 1 and $\sqrt{2}$ we have confirmed up to the 11th decimal digit; if this is proven, then a truly new field of analysis stands before us [56, X.2, p. 43]."

Expressed as a formula, the discovery, which even today ranks as sensational, looks like this:

$$\frac{\pi}{\varpi} = \text{AGM}(\sqrt{2}, 1) \quad (= \quad 1.19814\ldots) \tag{7.13}$$

In this formula Gauss recognised the connection between two apparently non-overlapping areas, namely the AGM and the lemniscate functions, which in turn are closely related to the elliptic functions. He was also aware at the time of the significance of his discovery as his conclusion ("new field of analysis") suggests. This conclusion was fully confirmed, moreover, in the 19th century by the work of Riemann, Jacobi et al.

It is striking in the diary entry that Gauss had arrived at his findings not through mathematical deduction but through simple comparison of two numbers. Normally mathematicians are cautious about inferring a law simply from numeric coincidence. But this was quite different; Gauss seems to have suspected that this coincidence was no chance occurrence. This was by no means the only instance in which numeric results led Gauss to infer mathematical relationships, but it was without a doubt the most prominent.

The wording in his diary also reveals that Gauss did not yet have a proof, and this he developed only later. When exactly is not known, but it is certain that by 23 December 1799 he actually had two proofs, and in 1818 he even published a third.

The first and second proofs run over the comparison of the series expansions of the left and right sides of (7.13). The third proof is the shortest and the most elegant [56, III, pp. 352, 353]. Gauss examined the complete elliptic integral of the first kind,

$$I(a,b) = \int_0^{\pi/2} \frac{d\theta}{\sqrt{a^2 \cos^2 \theta + b^2 \sin^2 \theta}}$$

of which the lemniscate perimeter is a special case: $\varpi = 2I(\sqrt{2},1)$. He then showed that this integral remains constant if a and b are iterated in accordance with the AGM rule (alg 7.2): $I(a,b) = I(a_1,b_1) = I(a_2,b_2) = \cdots$. As a_k and b_k move towards the same limit value $\mathrm{AGM}(a,b)$, $I(a,b) = \pi/(2 \cdot \mathrm{AGM}(a,b))$ applies, which proves the relation (7.13).

It is only a few steps from this proof to the π formula (7.1). In modern accounts, for example as presented by Nick Lord [81], the path from there takes one over the following steps which are not difficult and were not new ground in Gauss's time. The missing pieces were, firstly, the auxiliary integral $L(a,b) = \int_0^{\pi/2} \cos^2 \theta d\theta / \sqrt{a^2 \cos^2 \theta + b^2 \sin^2 \theta}$, for which $L(a,b) + L(b,a) = I(a,b)$ holds true; secondly, the relationship mentioned above that was discovered by Leonhard Euler(7.11) of $L(\sqrt{2},1) \cdot I(\sqrt{2},1) = \pi/4$; and, thirdly, the relationship $c_0^2 L(a,b) = (c_0^2 - S)I(a,b)$ with $S = \sum_{k=0}^\infty 2^{k-1} c_k^2$, which is obtained from $c_0^2 L(b,a) - 2c_1^2 L(b_1,a_1) = c_0^2 I(a,b)/2$. If the argument pair $(\sqrt{2},1)$ is converted to the pair $(1, 1/\sqrt{2})$, a more elegant version of the result is obtained. And there you have the formula (7.1) and the Gauss AGM algorithm.

The original

It took us a long while to uncover the original formula of Gauss which forms the basis of the Gauss AGM algorithm. It appears on page 6 of handbook 6, *Short essays from various fields of mathematics, begun in May 1809.*

This little work is in the possession of the Niedersächsische Staats- und Universitätsbibliothek Göttingen (Lower Saxon State and University Library Göttingen) (shelf mark: Cod. Ms. C. Fr. Gauss, Handbuch 6). When we enquired after it, it turned out to be in a pitiful condition, so that the manuscripts and rare books department first had to have the book restored before we were able to print the formula (with their blessing) for the first time as Gauss originally wrote it:

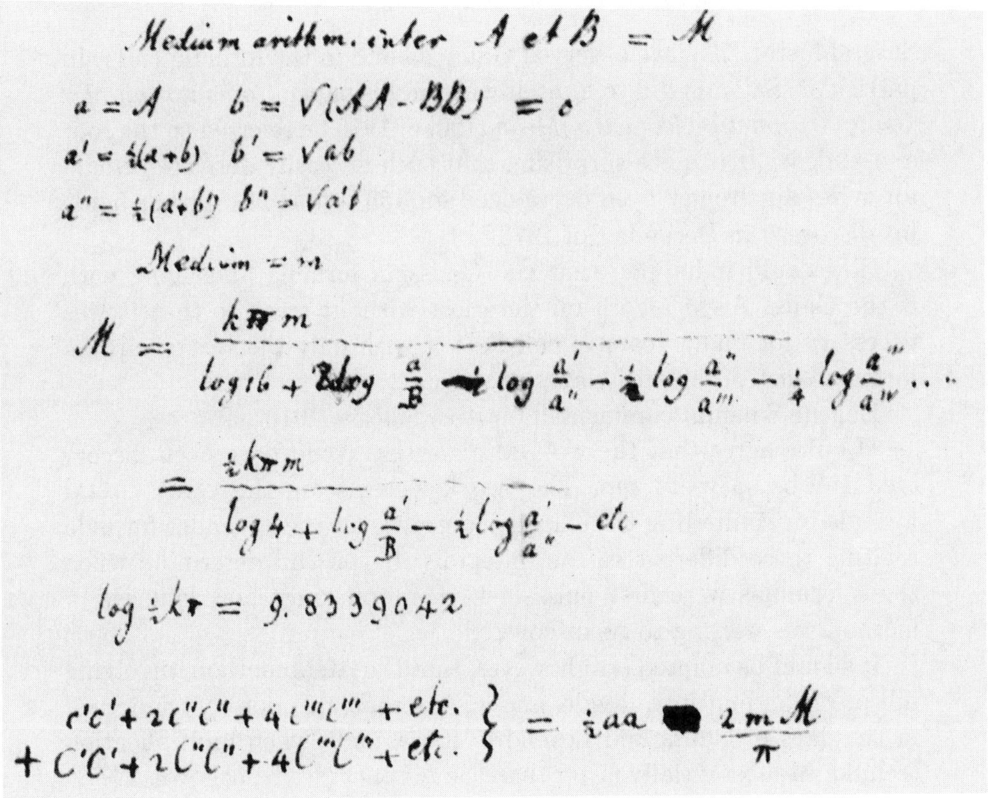

[handwritten facsimile:]

Medium arithm. inter A et $B = M$

$a = A \quad b = \sqrt{(AA-BB)} = c$
$a' = \tfrac{1}{2}(a+b) \quad b' = \sqrt{ab}$
$a'' = \tfrac{1}{2}(a'+b') \quad b'' = \sqrt{a'b'}$

Medium $= m$

$$M = \frac{k\pi m}{\log 16 + 2\log\tfrac{a}{B} - \tfrac{1}{2}\log\tfrac{a'}{a''} - \tfrac{1}{4}\log\tfrac{a''}{a'''} - \tfrac{1}{8}\log\tfrac{a'''}{a''''}\ldots}$$

$$= \frac{\tfrac{1}{2}k\pi m}{\log 4 + \log\tfrac{a'}{B} - \tfrac{1}{2}\log\tfrac{a'}{a''} - \text{etc.}}$$

$\log\tfrac{1}{2}k\pi = 9.8339042$

$\left.\begin{array}{l} c'c' + 2c''c'' + 4c'''c''' + \text{etc.} \\ + C'C' + 2C''C'' + 4C'''C''' + \text{etc.} \end{array}\right\} = \tfrac{1}{2}aa - \dfrac{2mM}{\pi}$

The formula in the last line

$$\left.\begin{array}{l} c'c' + 2c''c'' + 4c'''c''' + \text{etc.} \\ + C'C' + 2C''C'' + 4C'''C''' + \text{etc.} \end{array}\right\} = \frac{1}{2}aa - \frac{2mM}{\pi} \quad (7.14)$$

is what we were looking for! It is even more general than (7.1). It is only necessary to set $a = A = 1$ and $b = B = 1/\sqrt{2}$, and one obtains $m = M$ and $c^2 = C^2 = a^2 - b^2 = 1/2$. If instead of M and m one sets $AGM(1/\sqrt{2})$, and if each stroke is replaced by an index, for example if we write c_3 instead of c''', the Gauss formula is transformed into:

$$2\sum_{k=1}^{\infty} 2^{k-1} c_k^2 = \frac{1}{2} - \frac{2(AGM(1, 1/\sqrt{2}))^2}{\pi} \quad (7.15)$$

and from this the (7.1) is immediately derived.

How could it happen ...

Several historians have observed Gauss's hand in the formula (7.1). In particular, Salamin did so himself when in his paper *Computation of π Using Arithmetic-Geometric Mean* [100] of 1976 he referred to the root and added: "It is quite surprising that such an easily derived formula for π has apparently been overlooked for 155 years. The author made his discovery in December of 1973."

How could it happen that the Gaussian formula and along with it the Gauss AGM algorithm were lost without trace so that it was necessary for many years to calculate π with only the weaker arctan formulae instead of with Gauss's formula?

Eugene Salamin commented on it as follows [101]:

"I quite agree that the π-AGM algorithm could have been discovered 150 or so years ago. The two key items are the Gauss AGM formula for computing the elliptic integrals, and the Legendre formula relating these different elliptic integrals. If you can determine when these formulas were first published, then you know how long the π method was waiting to be discovered.

It should be pointed out, however, that the AGM method, involving full precision multiply, divide, and square root, was not very practical in the days of Gauss and Legendre. It was only when multiplication techniques substantially faster than the N^2 ones (see Knuth, vol. 2 [77]) were discovered, that the π AGM algorithm became more than just a theoretical formula."

That is all water under the bridge now. In 1976 the number of known digits of π stood at one million, this record having been established in 1973 using arctan series. After Brent und Salamin, this record was left in the dust. In only three years, computers were already chomping their way up to 30 million decimal places (see Chapter 13), all using programs that were based on the Gauss AGM algorithm.

8. Ramanujan and π

The story of Srinivasa Ramanujan, the greatest Indian mathematician of the modern era, could come from a sociology textbook on the making of geniuses. Impoverished youth, inadequate education, misunderstood by those around him, forced to adopt an independent line. All of a sudden someone spots the youth's brilliance, appreciates his visionary thinking, takes him under his wing, encourages him and elicits from him in the space of a few years an incredible wealth of discoveries. This creative period is followed by illness, homesickness and untimely death. In the ensuing period his work is largely lost without trace, known only to a few privileged persons, forgotten by the world at large. Then 60 years later he enjoys a kind of revival. A few people begin to get interested in him, enquire into his life and milieu, and suddenly this interest is kindled into a raging fire as a storm of excitement spreads through India and across the mathematically minded world. Jubilation culminates in celebrations, conferences and posthumous honours on the occasion of the centenary of Ramanujan's date of birth in 1987.

Ramanujan's extraordinary achievement was to tread new ground in mathematics despite having had little in the way of a formal mathematical education. As such, he stirred the creativity and imagination of the best brains and inspired them on to extraordinary discoveries of their own.

8.1 Ramanujan's series

A number of Ramanujan's discoveries are linked with the number π and the issue of how to calculate it. In 1914, at the age of 27, Ramanujan published an article entitled *Modular Equations and Approximations to π* [96]. This essay contains around 30 formulae for π, including the following [96, formula (44)]:

8. Ramanujan and π

$$\frac{1}{2\pi\sqrt{2}} = \frac{1103}{99^2} + \frac{27493}{99^6} \cdot \frac{1}{2} \cdot \frac{1 \cdot 3}{4^2} + \\ + \frac{53883}{99^{10}} \cdot \frac{1 \cdot 3}{2 \cdot 4} \cdot \frac{1 \cdot 3 \cdot 5 \cdot 7}{4^2 \cdot 8^2} + \cdots \quad (8.1)$$

Ramanujan himself commented that this series converges extremely rapidly[1]. In fact, every summand produces about 8 correct decimal places, as can be seen just from the first three terms:

$$0.1125\ldots \cdot 10^0 \qquad 0.273\ldots \cdot 10^{-8} \qquad 0.229\ldots \cdot 10^{-16}$$

In (8.1), Ramanujan had produced a π series which was significantly better than previous formulae, such as, for example, the arctan formula of John Machin, which moves towards π at the rate of only 1.4 decimal places per summand. Ramanujan's new series was thus the first one suitable for taking the calculation of π beyond the million mark, and in fact Gosper calculated 17 million places of π with this formula in 1985.

One might ask whether the great speed of convergence of the series (8.1) is not dissipated by the fact that what the formula calculates is not actually π but 1 *divided by* π, while it also includes another square root. But in fact this is not the case. Firstly, the calculation of the square root and of the inverse are operations which are only performed once and, secondly, procedures for calculating them which converge very rapidly, in fact quadratically, have been available since the time of Isaac Newton, i.e. for 400 years. The two terminal operations can thus be performed in the time it takes to perform a few long multiplication operations and add only insignificantly to the total time.

It is quite striking how it was in 1914 or even earlier that Ramanujan discovered the series (8.1), yet it was not used until 1985. The 70 years which passed between these two events were not a gestation period required for the series to mature, but the significant event which occurred in the interim was the invention of the computer. Without a computer, i.e. using paper and pencil only, it is not sensible to calculate an expression like the Ramanujan series over many terms. This and other products of Ramanujan's inventive genius therefore had to be put on ice until the tools needed to exploit them had been invented.

---

[1] In the literature this formula is generally presented in a slightly more compact form (see 13.49).

Ramanujan provided only sparse information on his series (8.1) and failed to furnish any proof. However, as is evident simply from the title of the article, the formula is related to modular equations.

It is rare to discover an account of modular equations that is readily comprehensible. Fortunately, however, the Borwein brothers have written such a description, which we may therefore assume to be authoritative [33, p. 112B]:

The Ramanujan π series (8.1) is based on a 58th order modular equation [11, p. 202]. Historians can only speculate as to how Ramanujan arrived at it. He himself says only that it was developed from theories "corresponding" to the standard theory of the theta and modular functions. We can be certain that it took Ramanujan a large number of attempts to arrive at manageable modular equations.

In his paper on π, as well as (8.1), Ramanujan lists another 16 series for $1/\pi$ which do not converge as well but some of which are quite spectacular for other reasons. One of these series is as follows [96, Formel (29)]:

$$\frac{1}{\pi} = \sum_{k=0}^{\infty} \binom{2n}{n}^3 \frac{42n+5}{2^{12n+4}} \qquad (8.2)$$

This series does not converge particularly well, producing only 1.8 decimal places per term. But it is possible with this series to calculate the second half of n binary places of $1/\pi$ without having to calculate the first half, because the denominators grow twice as quickly as the numerators. In this respect, Ramanujan had discovered here what was a kind of precursor to the BBP series discovered only quite recently (see Chapter 10), which permits the calculation of binary places in the "middle" of π.

8.2 Ramanujan's unusual biography

Srinivasa Ramanujan's short life has been portrayed in detail, the main impetus to this being the run-up to the centenary of his date of birth in 1987. The most comprehensive biography is that by Robert Kanigel [71]. The publisher of Ramanujan's notebooks, Bruce Berndt, has also collected together many biographical details [21], [22], [23]. In particular, he was able to interview Ramanujan's widow, S. Janaki Ramanujan, who survived her husband by more than 60 years.

8. Ramanujan and π

Srinivasa Aiyangar Ramanujan was born in the southern Indian city of Erode on 22 December 1887 in his mother's family's home. The family was very poor. After his birth, his mother returned to Kumbakonam, where her husband worked in a clothing business. Kumbakonam is about 155 miles to the south-west of Madras.

Ramanujan showed an early interest in and aptitude for mathematics. But, unlike many famous mathematicians, he had no mentor but acquired his mathematical knowledge entirely by himself. Up to the age of 25 it is likely that he only had access to five mathematical books. The first of these was S.L. Loney's *Trigonometry*, published in Cambridge in 1893, and this he worked his way through at the age of 12. The book discusses logarithms of complex variables, the calculation of π, the Gregory series and series sums, amongst other topics. He himself explored similar avenues later on.

At the age of 15, Ramanujan borrowed a work by George S. Carr, *A Synopsis of Elementary Results in Pure Mathematics* from the local public library. Carr's book was republished in 1970 under a different title [40]. This "synopsis" was in fact intended as an overview aimed at preparing university students for their finals in mathematics. The book lists 6,165 theorems, but without or with only very brief proofs. In any case (or notwithstanding), this book seems to have aroused Ramanujan's genius. When Ramanujan started documenting his mathematical discoveries in notebook form in 1904, Carr was evidently the model he followed.

Already at this time, aged 16, he was completely taken with mathematics and only wanted to study that subject. As a result he failed his examinations in the other subjects and was sent down. Further attempts to obtain a college education met with no success. "The Collage of Kumbakonam rejected the one great man they had ever possessed," Ramanujan's patron, Hardy, wrote bitterly [63, p. 7]. Consequently, from the age of 17 to 23, Ramanujan worked in complete isolation. He entrusted the results of his research to his notebooks but otherwise kept them to himself. It is very sad that he obtained no mathematical "feedback" at such an important period of his life.

In 1909 Ramanujan got married. His wife, who was only nine at the time, was able to provide some interesting details in 1984 (by which time she had been widowed for 64 years) [22]. In particular, she said Ramanujan was convinced that he would not have fallen ill in England if she had gone there with him.

Ramanujan's first mathematical publication appeared in 1911, and only from 1912 did his extraordinary mathematical genius begin to gain some recognition. Two English mathematicians, Sir Francis Spring and Sir Gilbert Walker, arranged a monthly stipend for him. After a long search, Ramanujan obtained employment at a harbour company in Madras in 1912. The proprietor of the company and a manager who was a mathematician himself took great interest in Ramanujan's work and encouraged him to inform some English mathematicians of his discoveries. His first attempts did not produce any encouraging results, but then in January 1913 he wrote to Godfrey H. Hardy, a very well-known English number theoretician at Trinity College, Cambridge University. Hardy was immediately convinced and invited Ramanujan to Cambridge. After having to face the objections of his caste and family (for orthodox Brahmins an ocean crossing was grounds for being made an outcast), Ramanujan sailed from India in 1914.

He began one of the most important collaborations in the history of mathematics. During his time in Cambridge Ramanujan published several articles, some of which are classified as fundamental. One of these is the above-mentioned article from 1914 on modular equations and approximations for π, which is where the series (8.1) comes from. This paper is the only one Ramanujan wrote about π, although additional findings relating to π are documented in his notebooks.

He wrote several articles jointly with Hardy. "The blend of Hardy's technical expertise and Ramanujan's raw brilliance produced an unequated collaboration" [33, p. 112A], commented the Borwein brothers.

After three years in Cambridge, Ramanujan succumbed to an unidentified illness in 1917. It was diagnosed as tuberculosis but it may simply have been vitamin deficiency due to poor nutrition during the First World War. After two years in sanatoriums and convalescence homes, he returned to his native country in 1919.

Soon after his return he was offered a chair at the Hindu University of Benares. Ramanujan was forced to decline for the moment due to his ill-health but planned to take up the offer when his health improved. But unfortunately he never recovered from his illness and died on 26 April 1920 at the age of 32 in Kumbakonam.

Ramanujan left 37 publications and a wealth of problems which he had published in the *Journal of the Indian Mathematical Society*.

There were also five notebooks and various unpublished essays and manuscripts plus 120 theorems in his letters to Hardy.

A simple list of the topics on which Ramanujan worked reads as follows [25, p. 644]: partitions, theta functions, statistical mechanics, Lie algebra, probability theory, modular forms, elliptic functions, complex multiplication, hypergeometric series, q series, asymptotic behaviour, beta integrals.

He also made some comparatively elementary discoveries. There is an anecdote about a taxi with the registration number 1729. It was in such a taxi that in around 1918 Hardy drove to a hospital to visit Ramanujan. On entering, Hardy mentioned this number, upon which Ramanujan commented, "1729 is a very interesting number, it is the smallest number expressible as a sum of two cubes in two different ways." Ramanujan was referring to the representations $1^3 + 12^3 = 9^3 + 10^3$, both of which produce 1729. In fact it is true that no smaller number possesses this characteristic. Ramanujan also discovered formulae for the construction of higher sums of powers and used them, for example, to ascertain the identity

$$4^4 + 6^4 + 8^4 + 9^4 + 14^4 = 15^4 \tag{8.3}$$

Ramanujan's skills at handling continued fractions are unsurpassed. Among his results, for example, we find – without proof, as was so often the case – the following continued fraction formulae which helped to convince Hardy of Ramanujan's genius. They are among the most aesthetically pleasing he left, but they are also some of the most difficult to prove [63, p. 8]:

$$\sqrt{\frac{5+\sqrt{5}}{2}} - \frac{\sqrt{5}+1}{2} = \cfrac{e^{-\frac{2\pi}{5}}}{1+\cfrac{e^{-2\pi}}{1+\cfrac{e^{-4\pi}}{1+\cfrac{e^{-6\pi}}{1+\cdots}}}} \tag{8.4}$$

$$\cfrac{\sqrt{5}}{1+\sqrt[5]{5^{\frac{3}{4}}\left(\frac{\sqrt{5}-1}{2}\right)^{\frac{5}{2}}-1}} - \frac{\sqrt{5}+1}{2} = \cfrac{e^{-\frac{2\pi\sqrt{5}}{5}}}{1+\cfrac{e^{-2\pi\sqrt{5}}}{1+\cfrac{e^{-4\pi\sqrt{5}}}{1+\cfrac{e^{-6\pi\sqrt{5}}}{1+\cdots}}}} \tag{8.5}$$

Hardy said of these and other formulae containing continued fractions [63, p. 9]: "I had never seen anything in the least like them before. A single look at them is enough to show that they could only be written down by a mathematician of the highest class. They [the formulae] must be true because, if they were not true, no one would have had the imagination to invent them."

Some of Ramanujan's findings have been lost, even though G.E. Andrews at least recovered Ramanujan's "missing notebook" in 1976. Some of it the University of Madras may have succeeded in losing. At any rate Mrs. Ramanujan said that shortly after her husband's death his former teacher had come and claimed all his documents for the university, but no one has succeeded in finding them there.

Ramanujan's *Collected Papers* appeared seven years after his death. In 1936, Hardy gave "twelve lectures on subjects suggested by [Ramanujan's] life and work", which were published in 1940. In the foreword, Hardy writes as follows about his protégé [63, p. 6]: "The real tragedy about Ramanujan was not his early death. It is of course a disaster that any great man should die young, but a mathematician is often old at thirty, and his death may be less of a catastrophe than it seems. Henrik Abel (1802–1829) died at twenty-six and, although he would no doubt have added a great deal more to mathematics, he could hardly have become a greater man. The tragedy of Ramanujan was not that he died young, but that, during his five unfortunate years, (between eighteen and twenty-three) - his genius was misdirected, sidetracked and to a certain extent distorted."

Several authors, including the Borwein brothers, have succeeded in recent years in rendering Ramanujan's work much easier to understand and more transparent. A significant contribution here was made (and is still being made) by Bruce Berndt, who is currently publishing Ramanujan's notebooks with a detailed commentary. Above all, Berndt is adding in the missing proofs (and in some cases disproofs) which Ramanujan omitted to write down or else wrote only incompletely. Four of the five notebooks have already been published [24]. For friends of π, notebooks 3 and 4 are the most interesting.

8.3 Impulses

The path which led Ramanujan to his discoveries about π has proved to be of great practical importance and very forward-looking. Several other researchers have continued down the same route.

The Borwein brothers have succeeded in deriving general formulae which enable any number of convergent series of the Ramanujan type to be generated (through fixing of one parameter). Further information will be found on our CD-ROM in the `arith` directory. Following this model, one of the authors (JA) generated the following series which produces a staggering 50 correct decimal places per term:

$$\frac{1}{\pi} = \frac{1}{\sqrt{-12\,J}} \sum_{n=0}^{\infty} \frac{(6n)!}{12^n\,(3n)!\,(n!)^3} \frac{A + nB}{J^n} \qquad (8.6)$$

where

$A := 52804190260809999654521\!85 +$
$\quad + 23614751784000701705688\!00 \sqrt{5} + 32\sqrt{5} \cdot$
$\quad \cdot \big(1089172855117117820046743621239520916038565601\!7 +$
$\quad + 4870929086578810225077338534541688721351255040 \sqrt{5}\big)^{1/2}$

$B := 654159204458052267524145750 +$
$\quad + 292548889855077669080467200 \sqrt{5} +$
$\quad + 209664 \sqrt{3110} \cdot$
$\quad \cdot \Big(6260208323789001636993322654444020882161 +$
$\quad + 2799650273060444296577206890718825190235 \sqrt{5}\Big)^{1/2}$

$J := -\Big[178977495886260\!20 + 8004116944887336 \sqrt{5} +$
$\quad + 108 \sqrt{5}\big(10985234579463550323713318473 +$
$\quad + 4912746253692362754607395912 \sqrt{5}\big)^{1/2}\Big]^3$

<small>Arndt [7], 1994</small>

Another spectacular but also practical "Ramanujan"-type series was developed by the Chudnovsky brothers. It goes like this:

$$\frac{1}{\pi} = \frac{12}{\sqrt{640320^3}} \sum_{k=0}^{\infty} (-1)^k \frac{(6k)!}{(k!)^3 (3k)!} \frac{13591409 + 545140134k}{(640320^3)^k} \quad (8.7)$$

D. and G. Chudnovsky, 1987

It was with this series, which produces 15 decimal places per series term, that the Chudnovskys calculated π to over 8 billion decimal places in 1996. The series is used in several computer algebra systems, e.g. in *Mathematica*, to calculate π, and is thus a very useful series; it would also be the ideal candidate for anyone seeking to divide up the task of calculating π on the Internet because it can be adapted very well to the binsplit algorithm (see Chapter 15 on page 215ff.).

In the more recent high-performance calculations of π, Ramanujan-type series have been replaced by other methods, notably by iterative algorithms developed by the Borweins.
One of these is this "quintic" algorithm:

Algorithm (8.8) Borwein, quintic (alg 8.8)

Initialise:
$$s_0 := 5(\sqrt{5} - 2)$$
$$a_0 := 1/2$$

Iterate:
$$s_{k+1} := (25)/(s_k(z + x/z + 1)^2) \xrightarrow{5} 1$$
where
$$x := \frac{5}{s_k} - 1$$
$$y := (x - 1)^2 + 7$$
$$z := \left(\frac{x}{2}\left(y + \sqrt{y^2 - 4x^3}\right)\right)^{1/5}$$
$$a_{k+1} = s_k^2 a_k - 5^k \left(\frac{s_k^2 - 5}{2} + \sqrt{s_k(s_k^2 - 2s_k + 5)}\right) \xrightarrow{5} \frac{1}{\pi}$$

Then:
$$a_n - \frac{1}{\pi} < 16 \cdot 5^n e^{-\pi 5^n}$$

J. and P. Borwein, 1987 [33, p. 115]

a_k produces 5th order convergence on $1/\pi$, i.e. on every iteration the number of decimal places found undergoes a fivefold increase.

Such algorithms may not be found in Ramanujan's writings, but they are closely related to Ramanujan's analysis. For example, this special algorithm is based on a 5th order modular equation (inside the relation s_k) whose origin can be traced back to Ramanujan. Thus one can say with some justification that Ramanujan's hand is still to be found in π calculations today.

9. The Borweins and π

If any one or two figures stand out from the π research of recent years, then it must be Peter and Jonathan Borwein. They have developed virtually all the high-performance algorithms on which π computations are based today.

The brothers were born in St. Andrews, Scotland, Jonathan in 1951 and Peter in 1953. Their parents emigrated to Canada, where David Borwein was for many years head of the mathematics institute at the University of Western Ontario and is now a professor emeritus. It was "natural" that the brothers should study mathematics, and Jonathan obtained his PhD in 1974, Peter in 1979.

After completing their studies, they worked – together – at Dalhousie University in Halifax, the capital city of the east Canadian province of Nova Scotia. There they both became professors as well. Then in 1992 Jonathan moved from the east to the west of Canada, to be followed a year later by Peter. Today they both teach and research at the Simon Fraser University in Burnaby, British Columbia. There they head the Center for Experimental & Constructive Mathematics (CECM), which they have had a free hand at establishing.

The Borweins' academic work is already very considerable. Peter Borwein's list of publications includes five books and almost 100 articles just over a period of 14 years. Jonathan Borwein has published a similar volume of papers.

π is not the only research field in which the Borweins are active. In fact, less than 20% of their publications include π in the title. Nevertheless π is the main topic with which their names are currently linked. The main reason for this is that since 1984 the brothers have developed a large number of high-performance algorithms for the calculation of π which have been immediately implemented and executed by various mathematicians, such as Bailey and Kanada. The Borweins made their debut in the π field in 1984 with the following quadratically convergent algorithm [30], which they derived from the π algorithm of Gauss, Salamin and Brent (see Chapter 7):

Algorithm (9.1) π AGM derived, quadratic (alg 9.1)

Initialise:
$$a_0 := \sqrt{2}$$
$$b_0 := 0$$
$$p_0 := 2 + \sqrt{2}$$

Iterate ($k = 0, 1, 2, \ldots$):
$$a_{k+1} := \tfrac{1}{2}\left(\sqrt{a_k} + \tfrac{1}{\sqrt{a_k}}\right)$$
$$b_{k+1} := \sqrt{a_k}\left(\tfrac{b_k+1}{b_k+a_k}\right)$$
$$p_{k+1} := p_k b_{k+1}\left(\tfrac{1+a_{k+1}}{1+b_{k+1}}\right) \xrightarrow{2} \pi+$$

Then converge the p_k's quadratically to $\pi+$, where:
$$p_k - \pi < 10^{2^{k+1}} \quad (k \geq 2)$$

J. and P. Borwein, 1987 [30, p. 360]

But this algorithm was only the starting point for many other algorithms of the iterative type, including some which converge even better. Among these, the following quartic algorithm made their names especially well known, as since its publication in 1987 it has been used in most subsequent world record calculations, notably in those achieved by Yasumasa Kanada [32, p. 170]:

Algorithm (9.2) Borwein, quartic (alg 9.2)

Initialise:
$$y_0 := \sqrt{2} - 1$$
$$a_0 := 6 - 4\sqrt{2}$$

Iterate ($k = 0, 1, 2, \ldots$):
$$y_{k+1} := \tfrac{1-\sqrt[4]{1-y_k^4}}{1+\sqrt[4]{1-y_k^4}}$$
$$a_{k+1} := a_k(1+y_{k+1})^4 - 2^{2k+3}y_{k+1}(1+y_{k+1}+y_{k+1}^2) \xrightarrow{4} \tfrac{1}{\pi}$$

Then converge the a_k quartically to $1/\pi$

The error is:
$$0 \; a_k - \tfrac{1}{\pi} \leq 16 \cdot 4^k \cdot 2e^{-4^k 2\pi}$$

J. and P. Borwein, 1987 [32, p. 170]

This algorithm is expressed in a more general form together with many other algorithms in the π textbook Jonathan and Peter Borwein published in 1987 under the title [32]

Pi and the AGM
A Study in Analytic Number Theory and Computational Complexity

This important book (a *must* for all π enthusiasts who dare to try their hand at substantial mathematical fare) covers all aspects of π research, not just the rapid computation of π to many decimal places. For example, the authors go back to the 19th century, to the analysis of the time, elliptic integrals and functions, theta and modular functions, Lagrange (1736–1813), Legendre (1752–1833), Gauss (1777–1855), Jacobi (1804–1851) and Ramanujan (1887–1920).

The whole book is permeated with the respect and admiration of the two young writers for the achievements of those on whose shoulders they stand, who researched the subject before them and provided the groundwork for their own discoveries. The voluminous list of references are a testament to this, as is the fact that they dedicate the book to their father, David Borwein.

In addition to the high-performance algorithms themselves, for which they have become famous, the Borweins explain their general structure. They prove, for example that the Ramanujan series (8.1) is the special case $n = 58$ of a much more general approach. And on the few pages of their Chapter 5, algorithms pour out one after the other – quadratic, cubic, quartic, quintic, septic and even nonic algorithms. The reader is invited to develop additional ones. Incredible. (You will find more on this subject on our CD-ROM in the directory `arith`).

Even after publication of their π book, the Borweins have remained true to the number π. In 1995 Peter Borwein and two other researchers came up with the spectacular discovery known as the Bailey-Borwein-Plouffe series (see Chapter 10). This series makes possible something one can scarcely imagine, namely that it might be possible to calculate individual (hexadecimal) digits right in the middle of π *without* having to calculate all the previous digits first. Just try and imagine the problem as a simple multiplication operation. From a product of 30-digit numbers, find the 29th digit under the condition that you may *not* calculate the digits either to the right or left of it. Sounds difficult indeed. Now suppose, for example, that you want to find the 123 billionth digit of π even though this digit is inextricably linked with its predecessor digits in a gigantic tree. Yet it is possible.

Just recently a new book by Jonathan and Peter Borwein (together with Lennart Berggren) has been published [20],

Pi: A Source Book

This book contains 70 important original articles on the subject of π in facsimile form, for example the historically important treatises by Lambert on the irrationality of π (1761) and by Lindemann on the transcendence of π (1882). However, the book also includes some important articles by the Borweins themselves. This excellent and useful book confirms the historic dimension which they see as providing a setting to their own work.

The Borwein brothers' ground-breaking work has not only ensured that the number of known π digits has increased enormously, but above all we owe it to the Borweins that π research is today held in high esteem - even among mathematicians, of whom not a few believed that the whole topic had been laid to rest since Lindemann's proof of the transcendence of π.

On the basis of the results achieved by the Borweins, it has even been suggested that separate branches of mathematics should be established for π and e, which might then be known respectively as π *mathematics* and e *mathematics*. e mathematics would be linear, explicit, open to generalisation, easily approachable, highly algebraic and would lead via the exponential function to topics like one-parameter subgroups, Lie algebra and group representations. By contrast, π mathematics would be non-linear, chthonic, barely generalisable, highly analytical and, via modular functions and Ramanujan identities, would have a significant impact on function theory, number theory and combinational logic [121].

Who knows, perhaps one day there will be an academic degree of Dr. math. π. Anyone holding such a degree would have Jonathan and Peter Borwein to thank.

10. The BBP Algorithm

The title of the invitation sounded merely interesting:

> On the nth digit of a transcendental number OR The 10 billionth hexadecimal digit of π is '9'.

However, a sensation was concealed in the text which followed:

> "Up to now, it was generally believed that to compute the nth digit of a transcendental number like π was as difficult as calculating π itself. *We will show that this is not true* ...
> We will give algorithms for this ... These algorithms can be easily implemented. do not need multiple precision arithmetic, require virtually no memory, and feature run times that scale nearly linearly with the order of the digit desired. They make it feasible to compute, for example, the billionth binary digit of $\log 2$ or π on a modest work station in a few days of run time..."

Such was the wording of the title and abstract with which invitations were issued on 30 September 1995 to attend a colloquium at the Simon Fraser University in Burnaby, Canada. The invitations had been issued by three well-known π mathematicians, David Bailey, Peter Borwein and Simon Plouffe.

Two weeks later at the colloquium the three speakers presented the algorithms they had announced, gave the necessary proofs for them and with this overturned the old adage, he who seeks to climb to a mountain peak (or to calculate the furthermost digits of π) must climb the mountain by foot (or start from the initial "3").

It was a new formula which destroyed an old "paradigm":

$$\pi = \sum_{n=0}^{\infty} \frac{1}{16^n} \left(\frac{4}{8n+1} - \frac{2}{8n+4} - \frac{1}{8n+5} - \frac{1}{8n+6} \right) \qquad (10.1)$$

With this series it thus becomes possible to calculate any hexadecimal place *within* π. Even better, the series enables the next digits *from* any given hexadecimal place of π, for instance from the last known digit, to be added piece by piece without having to calculate a single prior position. The distinguishing term in the formula is the 16^n in the denominator of every series term.

We will shortly examine the mathematical procedure of the "BBP series" (as it has since become known). But first let us consider the question of how the formula came about.

Certainly not by chance, even if luck played some part in the discovery. All three parties are established mathematicians who have been working with the number π for a considerable time. The mathematical background against which the BBP series was developed has been thoroughly researched. Yet the series was *not* discovered through mathematical deduction or inference. Instead, the researchers used a tool called Computer Algebra System and a particular procedure called the "PSQL algorithm" to generate their series. They themselves write [10] that they found their formula (10.1) "through a combination of inspired testing and extensive searching".

Computer Algebra is a scientific area which is concerned with methods for the solution of mathematically formulated problems using symbolic algorithms. Its origin is sometimes attributed to Ada Augusta, Countess of Lovelace[1] (1815–1852), who may have been the first person to remark that computers are also suited for handling *symbolic data* (such as formulae). Computer Algebra Systems, of which the best known are *MuPAD*, *Maple* or *Mathematica*, normally work with mathematical symbols rather than with numbers. One of their use is for checking the derivations of equations, but they are also used to identify new relationships, as in the present context or to perform high-precision calculations.

The PSQL algorithm is used to find integer relationships between real numbers. A vector of real numbers $(x_1, x_2, \ldots, x_n)$ is entered. The algorithm then looks for a vector of integers $(a_1, a_2, \ldots, a_n)$, which do not all $= 0$, such that the relation $a_1 x_1 + a_2 x_2 + \cdots + a_n x_n = 0$ holds true. It is easy to see why this algorithm was a material factor in discovering the BBP series. The researchers had established that the classic series for $\ln 2 = \sum_{n=1}^{\infty} \frac{1}{n \cdot 2^n}$ can be used to calculate any binary digits of $\ln 2$. They therefore wondered whether there might be a similar

---

[1] The programming language ADA is also named after the Countess.

series for π. After many failed attempts, they finally fed the PSQL algorithm with $x_1 = \pi$, $x_2 = \sum_{n=1}^{\infty} \frac{1}{(8n+1)16^n}, \ldots, x_8 = \sum_{n=1}^{\infty} \frac{1}{(8n+7)16^n}$. When the algorithm then churned out the vector $(1, -4, 0, 0, 2, 1, 1, 0)$, the BBP series was discovered [58].

In this way the series (10.1) was the brainchild not of a human but a computer. Nevertheless, it still needs a normal proof, which is not very difficult [13].

First it is necessary to establish that for every $k < 8$ the following is true:

$$\int_0^{1/\sqrt{2}} \frac{x^{k-1}}{1-x^8} dx = \int_0^{1/\sqrt{2}} \sum_{n=0}^{\infty} x^{k-1+8n} dx$$

$$= \frac{1}{2^{k/2}} \sum_{n=0}^{\infty} \frac{1}{16^n(8n+k)} \quad (10.2)$$

Then one can write:

$$\sum_{n=0}^{\infty} \frac{1}{16^n} \left(\frac{4}{8n+1} - \frac{2}{8n+4} - \frac{1}{8n+5} - \frac{1}{8n+6} \right) =$$

$$= \int_0^{1/\sqrt{2}} \frac{4\sqrt{2} - 8x^3 - 4\sqrt{2}x^4 - 8x^5}{1-x^8} dx \quad (10.3)$$

If one inserts $y = x\sqrt{2}$, one obtains

$$\int_0^1 \frac{16y - 16}{y^4 - 2y^3 + 4y - 4} dy$$

$$= \int_0^1 \frac{4y}{y^2 - 2} dy - \int_0^1 \frac{4y - 8}{y^2 - 2y + 2} dy = \pi \quad (10.4)$$

where the fraction in the integral on the left has been partly reduced.

The BBP series could certainly also be used to calculate π from scratch, but it is not particularly well suited for such a computation, so it will never replace algorithms such as, for example, the Gauss AGM algorithm, which have been known for longer. The strength of the formula (10.1) clearly lies in its ability to calculate individual digits "anywhere" in π. The computing plan for this is not at all difficult and moreover it contains a special treat.

During evaluation of the BBP series, one calculates each of the four summands $S_1 = \sum_{n=0}^{\infty} \frac{1}{16^n(8n+1)}$, S_2, S_3, S_4 separately. Let us consider, for the sake of example, S_1.

To obtain the sequence of hexadecimal digits of S_1 which begin at a given position $p = d+1$ ($p = 1, 2, \ldots$), S_1 is shifted towards the left by d hexadecimals places. The part shifted beyond the hexadecimal point is discarded. The digits are then obtained from the part of the product $16^d \cdot S_1$ which appears after the hexadecimal point. When it comes to the divisions by $8n + 1$, once again only the places after the hexadecimal point are of interest, so that only the remainder of these divisions ever have to be calculated, as a result of which the numbers carried along are enormously reduced. In this way, the calculation of S_1 is simplified in the following way:

$$\text{Decimal portion of } 16^d S_1 =$$
$$= \sum_{n=0}^{\infty} \frac{16^{d-n}}{8n+1} \bmod 1$$
$$= \sum_{n=0}^{d-1} \frac{16^{d-n} \bmod (8n+1)}{8n+1} \bmod 1 + \sum_{n=d}^{\infty} \frac{1}{16^{n-d}} \frac{1}{8n+1} \qquad (10.5)$$

Here, reduction through the modulo function is used wherever appropriate, so that only the remainder of each division is considered. The summand S_1 is now represented by two part-sums. In the first of these, the $d-n$'s are positive so that the powers 16^{d-n} are in the nominator and the series terms are > 1. In the second part-sum, on the other hand, the $d-n$'s are negative and all the series terms are < 1. The reduction to remainders is only necessary and useful for the first part-sum.

10.1 Binary modulo exponentiation

The treat we referred to above occurs during calculation of the numerators of the first part-sum $16^{d-n} \bmod (8n+1)$. If the modulo operation in the first part-sum is initially ignored, then we have the most effective calculation possible for a power of the form x^k. There is an old procedure for this which is described by Knuth [77, p. 461] as follows:

Suppose, for example, that we need to compute x^{16}; we could simply start with x and multiply by x fifteen times. But it is possible to obtain the same answer with only four multiplications, if we repeatedly take the square of each partial result, successively forming x^2, x^4, x^8, x^{16}.

The same idea applies, in general, to any value of k, in the following way. Write k in the binary number system (suppressing zeros at the left). Then replace each "1" by the pair of letters SX, replace each "0" by S, and cross off the "SX" that now appears at the left. The result is a rule for computing x^n, if S is interpreted as the operation of *squaring*, and if X is interpreted as the operation of *multiplying by x*.

For example, if $k = 23$, its binary representation is 10111; so we form the sequence SX S SX SX SX and remove the leading SX to obtain the rule SSXSXSX. This rule states that we should "square, square, multiply by x, square, multiply by x, square, and multiply by x; in other words, we should successively compute x^2, x^4, x^5, x^{10}, x^{11}, x^{22}, x^{23}.

According to Knuth this "binary exponentiation" was already known in India over 2000 years ago. It is recorded in 200 BC in India in a work called "Chandah-sûtra" [77, p. 461]. Knuth notes that this information was evidently confined to India and outside of India there was still no mention of the method 1000 years later.

In the problem we are about to consider, a small modification should be made. As we are only interested in the places after the hexadecimal point and integer elements are negligible in the calculation, we only need to calculate the remainders of each division and can forget the quotients. This makes all the numbers a lot smaller. Thus on every iteration, a modulo operation is performed with the (next) denominator $c = 8n+1$. The result is "binary modulo exponentiation". This can be easily translated into an algorithm:

Algorithm (10.6) Binary modulo exponentiation (alg 10.6)

To calculate $r = b^k \mod c$, let $r := b \mod c$ and t be the value of the greatest power of 2 which is not bigger than k. Set $k := k - t$. Then:

while ($t > 1$)
{
    $t := t/2$;
    $r := r^2 \mod c$
    if $(k \geq t)$
    {
        $k := k - t$
        $r := b \cdot r \mod c$
    }
}

At the end of this calculation, r is given the desired value $b^k \mod c$. The algorithm is executed entirely with positive integers which do not become greater than $(c-1)^2$. A numeric example from [13] goes as follows. During the calculation of $3^{49} \mod 400$, r successively assumes the values 3, 9, 27, 329, 241, 81, 161 and 83, and in fact $3^{49} = 239,299,329,230,617,529,590,083$, so that 83 is the correct result.

Calculation of the rest of the formula (10.1) is very simple. The summands calculated through binary modulo exponentiation have to be divided by $8n+1$ and added up. Once again we are only interested in the portions which come after the hexadecimal point. The procedure is then expanded to the summands S_2 to S_4. The sum $4S_1 - 2S_2 - S_3 - S_4$ is then calculated, and its first place after the hexadecimal point is the desired hexadecimal digit of π.

As you can see, the BBP series is surprisingly simple. Neither the formula (10.1) nor its proof or implementation are particularly subtle or complicated. It is precisely this which makes the algorithm so interesting.

Even if π is the main point of interest, as is so often the case, especially in this book, it requires only a minor foreshortening of the results of Bailey, Borwein and Plouffe to narrow it to π. In fact it goes a lot further. The three researchers have not only found additional formulae of the above type for π, π^2 etc., but they have also discovered such formulae for other "polylogarithmic" constants, such as $\log 2$.

Their paper on the subject [10] presents these results in detail as well. We recommend most strongly that any reader so inclined should look at the article, which can be easily downloaded from the Internet.

10.2 A C program on the BBP series

Below we present an ANSI C program for calculating the pth hexadecimal place after the hexadecimal point of π. p can be up to around 536 million. The program uses the BBP series (10.1) of Bailey, Borwein and Plouffe.

It is almost the same as the program which David Bailey has placed on the Internet[2].

In the program, the central operation of the binary modulo exponentiation is performed in the function `expm()`. After the description, the source code should be easy to understand.

The function `series()` is used to compute each of the 4 summands in the BBP series. As explained above, each of these summands has two part-sums. The first consists of the first $d = p - 1$ terms with numerators > 1, and the second consists of the succeeding terms with numerators $= 1$. The portion of the terms in the first summand which comes after the hexadecimal point is calculated using binary modulo exponentiation; the terms in the second summand are all < 1 and are determined in the normal way using floating point division. From the second summand, only so many terms are used until, taking roundings into account, the result is accurate to 8 hexadecimal places, for the program outputs not just the pth place but also the next 7 places. To achieve a level of accuracy of 8 hexadecimal places after the hexadecimal point one needs around 42 binary places, corresponding to a decimal accuracy of around 10^{-13}, which is in fact the value of `eps` in the program.

In `main()` the user parameter p is read in, calculation of the four summands is triggered and finally the post-hexadecimal point portion from position p is printed as a hexadecimal number.

The time it takes to run the program is a little more than of a linear order. To compute the millionth hexadecimal place of π takes 48 seconds on our Pentium 400 platform, while the 10 millionth place takes 580 seconds.

---

[2] http://www.cecm.sfu.ca/personal/pborwein

It should be possible to compile the program with any ANSI C compiler.

```
/***********************************************************
 *
 * This ANSI C-Program computes the p-th hexadecimal digit
 * of pi  (plus the following 7 digits).
 *
 * p comes from the command line.
 * 0 <= p <= 2^29 (approx 536 million).
 * p=1 means the first digit after the radix point.
 *
 * The program uses the BBP-algorithm of
 * D. Bailey, P. Borwein und S. Plouffe.
 * Its central function is expm() which performs the
 * "binary modulo exponentation".
 * After D. Bailey 960429
 *
 ***********************************************************/

#include <stdio.h>
#include <stdlib.h>
#include <math.h>
#include <time.h>
#include <float.h>

/* Limits:
      d_max  = p_max - 1
      ak_max = 8 * d_max + 6
      r_max  = ak_max - 1
      r_max^2 < 2^(LDBL_MANT_DIG).
      With mantissa length = 2m (usually 2m = 64):
      r_max  <= 2^m-1
      ak_max <= 2^m
      d_max  < 2^(m-3)
      p_max  < 2^(m-3)
*/
#define p_max   pow(2, LDBL_MANT_DIG*0.5 - 3)

/***********************************************************
 *
 *         expm(long n, double ak)
 *
 * Computes 16^n mod ak using binary modulo exponentation.
 *
 ***********************************************************/
double expm(long n, double ak)
{
   long double r = 16.0;
   long nt;

   if (ak == 1) return 0.;
   if (n == 0)  return 1;
   if (n == 1)  return fmod(16.0, ak);

   /* nt: largest power of 2 not greater han n/2. */
   for (nt=1; nt <= n; nt <<=1)
      ;
```

```c
    nt >>= 2;

    /*  Binary exponentiation modulo ak. */
    do
    {
        r = fmodl(r*r, ak);
        if ((n & nt) != 0)
            r = fmodl(16.0*r, ak);
        nt >>= 1;
    } while (nt != 0);
    return r;
}
/***********************************************************
 *
 *              series(double m, long d)
 *
 *  computes sum_k {16^(d-k)mod(8k+m)}/(8k+m),
 *                      (k=0, 1, 2, ..., d-1)
 *
 ***********************************************************/
double series(double m, long d)
{
    long    k;
    double ak = m;
    double s = 0., t = 1., x;

    /*  Sum of terms 0..d-1 */
    for (k = 0; k < d; k++)
    {
        x = expm (d-k, ak);
        s += x / ak;
        s = fmod(s, 1.0);
        ak += 8.0;
    }

    /* Some additional terms for 8 hex digit accurracy */
    #define eps 0.25e-12 /* = 16^{-10} / 4 */

    while ( (x=t/ak) > eps)
    {
        s += x;
        t /= 16.0;
        ak += 8.0;
    }
    return s;
}

/***********************************************************
 *
 *              main()
 *
 ***********************************************************/
main(int argc, char *argv[])
{
    time_t t_beg, t_end;
    double s;
    double p = 1000.0;                  /* default fuer p */
```

```
/* Get p from the command line */
if (argc > 1)   p = strtod(argv[1], NULL);
if (p < 1)      p = 1;
if (p > p_max)  p = p_max;

printf("Hex digits %.0f to %.0f of pi: ", p, p+7);
t_beg=clock();
s =   4*series (1, p-1)      /*  4, 8i+1 */
    - 2*series (4, p-1)      /* -2, 8i+4 */
    -   series (5, p-1)      /* -1, 8i+5 */
    -   series (6, p-1);     /* -1, 8i+6 */
s += 4;               /* ensure s >= 0 */

t_end=clock();
printf("%081X\n", (unsigned long)(s*pow(2, 32)));
printf("Elapsed time was: %.1f sec.\n",
                   1.0*(t_end-t_beg)/CLOCKS_PER_SEC);
return 0;
}
```

10.3 Refinements

Bailey, Borwein and Plouffe had hardly announced their "shocking formula" (10.1) when other researchers sat down at their computers and attempted to also find beautiful "BBP-like" series. In particular, BBP formulae were found for other constants, for example, for π^2 and $\log 7$, as well as simpler or more effective formulae for π itself (to view some examples, see our collection of formulae on page 227).

The research duo Viktor Adamchik and Stan Wagon win the prize for the simplest of all the BBP formulae for π:

$$\pi = \sum_{n=0}^{\infty} \frac{(-1)^n}{4^n} \left(\frac{2}{4n+1} + \frac{2}{4n+2} + \frac{1}{4n+3} \right) \qquad (10.7)$$

Adamchik, Wagon, 1997 [2]

This formula consists of only 3 terms, i.e. one term fewer than the original BBP series. But it is not only shorter.

Wagon and Adamchik have determined with this series [1] that there is a route to its derivation which simultaneously includes the proof of the series. With (10.7) one no longer needs to convert the summands found by the computer into integrals and then resolve them, as shown above, but instead one can to a certain extent lean back and leave the computer to manipulate identities. The authors needed "only" one *idea*.

As such an idea, they used the expression:

$$\sum_{n=0}^{\infty} \frac{(-1)^n}{4^n} \left(\frac{a_1}{4n+1} + \frac{a_2}{4n+2} + \frac{a_3}{4n+3} + \frac{a_4}{4n+4} \right)$$

They entered this into their computer algebra system *Mathematica 3.0* which then produced the following transformed expression:

$$\frac{1}{8}(2(4(a_2 \operatorname{arccot} 2 - a_4 \log 4 + a_4 \log 5 + a_3(\pi/4 + \operatorname{arccot} 3 -$$
$$- (\log 25)/4))) + a_1(\pi + 4 \operatorname{arccot} 3 + \log 25))$$

At this point the authors told the computer to perform a few simple substitutions, e.g. the substitution $\pi = \arctan 1 + \arctan 2 + \arctan 3$, following which it produced this still further simplified output:

$$\frac{a_2 \pi}{2} + \left(\frac{a_1}{2} - a_2 + a_3 \right) \arctan 2 - 2a_4 \log 2 + \left(\frac{a_1}{4} - \frac{a_3}{2} + a_4 \right) \log 5$$

This took care of the most irksome work. Wagon and Adamchik now needed only to look for such values for $a_1, a_2, \ldots a_4$, with which the coefficients of the second to fourth terms become $= 0$, while the first coefficient becomes $= 1$. This solution they would of course have also been able to find in 10 seconds by hand, but they got the computer to do this for them too, and for this purpose they entered

Solve $\left[\left\{ \dfrac{a_2}{2} == 1, \dfrac{a_1}{2} - a_2 + a_3 == 0, a_4 == 0, \dfrac{a_1}{4} - \dfrac{a_3}{2} + a_4 == 0 \right\} \right]$

They could hear their computer deliberating, and then the system delivered quite simply:

$$\{\{a_2 \to 2, a_1 \to 2, a_3 \to 1, a_4 \to 0\}\}$$

And with that the above simple π formula (10.7) was born and proven at the same time! Child's play, really...

Short and instructive though the formula (10.7) is, it is not particularly effective. Fabrice Bellard overcame this shortcoming with the following series:

$$\pi = \frac{1}{64} \sum_{n=0}^{\infty} \frac{(-1)^n}{1024^n} \left(-\frac{32}{4n+1} - \frac{1}{4n+3} + \frac{256}{10n+1} - \right.$$
$$\left. - \frac{64}{10n+3} - \frac{4}{10n+5} - \frac{4}{10n+7} + \frac{1}{10n+9} \right) \qquad (10.8)$$

Obviously (10.8) is quicker than the two previously cited BBP series for π. Our own measurements confirm the 43% saving which Bellard claims for this series compared with (10.1). This improvement is based on the fact that Bellard's formula calculates to the higher base 1024, compared with the base 16 of the BBP series, so that he needs fewer terms for the same search item. This formula enables one to get further into π with the same amount of effort.

Bellard's series is currently the favourite amongst π digit hunters. With it, Bellard and most recently the student Colin Percival have broken several records, and as of September 2000 the latest record was the 10^{15}th binary place, a 0_2 (see page 203).

The BBP series and the many variants which have subsequently been developed unfortunately can only produce *hexadecimal* (or binary, octal etc.) places, and this is at least somewhat exotic. It would be nicer if we had a formula which could produce any *decimal place* of π.

Now, for some time we have known of a constant whose hindividual decimal places can be calculated with a BBP-like series. This constant is $\ln(0.9)$ and its related series goes like this:

$$\ln\left(\frac{9}{10}\right) = -\sum_{n=1}^{\infty} \frac{1}{10^n} \frac{1}{n} \qquad (10.9)$$

This series raises hopes that for π too a series will one day be discovered whose denominators contain a power of 10 rather than a power of 2, so that we can then explore decimal, and hence familiar terrain.

Unfortunately, this is not very likely. One of the distinguishing features of the BBP formulae for π is the fact that they contain arctan expressions which are rational multiples of π to base 16. (Unfortunately) it has meanwhile been shown that there are no arctan expressions with the base 10, so that the search for BBP formulae for decimal places is basically a waste of time. But perhaps someone will have a stroke of luck in some other place.

Simon Plouffe recently showed such a different path for the decimal place problem. He has specified an algorithm with whose aid individual digits of π can be found for any base. However, the procedure has the drawback that the time it takes to calculate is of cubic order and hence slower than the algorithms for calculating a decimal π from scratch.

But even this has already been improved on. The previously mentioned Fabrice Bellard has found out how to arrive at any individual digit of π in any base B in quadratic time. Here is his algorithm [19]:

Algorithm (10.10) Calculation of the nth digit of π (base B) (alg 10.10)

Set $N := \lfloor (n+\varepsilon)\log_2 B \rfloor$, where ε is a small integer (for example, 20) used to ensure accuracy.

$sum := 0$
for (every prime number a from $a = 3$ to $a < 2N$)
{
    $vmax := \lfloor \log(2N)/\log(a) \rfloor$
    $m := a^{vmax}$
    $v := 0$
    $s := 0$
    $A := 1$
    $b := 1$
    for $(k = 1, 2, ..., N)$
    {
        $b := \frac{k}{a^{v(n,k)}} \cdot b \bmod m$
        $A := \frac{2k-1}{a^{v(a,2k-1)}} \cdot A \bmod m$
        $v := v - v(a,k) - v(a, 2k-1)$
        if $(v > 0)$
        {
            $s := s + k \cdot b \cdot A^{-1} \cdot a^{vmax-v} \bmod m$
        }
    }
    $s := s \cdot B^{n-1} \bmod m$
    $sum := sum + \frac{a}{m} \bmod 1$
}

At the end of the algorithm, the desired decimal place stands at the beginning of the portion of sum which is after the base B point.

11. Arithmetic

Computation of π requires three things: a high-performance computer, an efficient algorithm and fast arithmetic. "Why fast arithmetic?" you ask. "Isn't that just the same thing as a high-performance computer?"

Yes and no. Naturally all computers can add up and multiply. This is their real domain, and high-performance computers do this too, albeit more quickly. But they can only work with short numbers, generally no longer than 32 or 80 bits, which is a long way short of what is required for the π problem. This means in effect that every π program has to assemble its long arithmetic operations out of the short operations which are all the computer offers.

11.1 Multiplication

The arithmetic operation which has by far the biggest impact on the length of time required is the multiplication of long numbers. A program for the calculation of π spends almost its entire run time performing either multiplication operations or else "higher" operations such as division or the calculation of square roots, which are actually implemented as a sequence of multiplications. By contrast, addition and subtraction play a relatively insignificant role.

What is the most effective way to calculate the product of two multiplicands? The answer to this question has been the same for hundreds of years, namely the same way as ever and as is still taught in school.

Anyone who learned his arithmetic before the era of the pocket calculator will know how school multiplication works: multiply the first multiplicand with each digit of the second multiplicand, write the product on a line underneath, then do the same for the other digits, displacing the results by one digit as appropriate, and then add all the lines of products together, taking into account any carryovers.

Because under the school method each digit of the first multiplicand is multiplied by each digit of the other multiplicand, if both multiplicands are N digits long, then a total of $N \cdot N = N^2$ individual multiplication operations must be performed. The time this takes rises as a function of N^2, so that it takes *four times* as long to multiply numbers which are only *twice* as long. This characteristic pattern means that if N is large, multiplication of the two numbers takes a huge amount of time. Even on a computer which can perform 10 million single multiplications per second, multiplication of two numbers each a million digits long will take more than a day to perform. In a high-precision computation with multiplicands that are considerably longer the time required for the necessary calculations is prohibitive.

But in fact there are faster ways of multiplying numbers, and it is not even difficult to understand how they work.

11.2 Karatsuba multiplication

The first way to accelerate the process is to use the *Karatsuba multiplication method*. Although this technique has only been known since the beginning of the 1960s, it is not clear who invented it. Evidently it was not invented by the Russian scientist Anatolii Karatsuba, after whom it is named, as in 1962 he only discovered a similar (and more complicated) method [77, p. 295]. Never mind.

Suppose we want to multiply together (in base 10) two $N = 2^k$-digit numbers u and v. (For the sake of simplicity we will assume here and in what follows that the two multiplicands in a multiplication are always of the same length and that their number of digits is an integral power of two. More complicated cases we will leave to the programmers.) Each of these numbers is divided into an upper and lower half, each of $N/2$ digits, and designated as u_1 and u_0, and v_1 and v_0, respectively, such that $u = 10^{N/2}u_1 + u_0$ and $v = 10^{N/2}v_1 + v_0$.

Under the school method, the product of u and v is calculated as follows:

$$uv = 10^N u_1 v_1 + 10^{N/2}(u_0 v_1 + u_1 v_0) + u_0 v_0 \qquad (11.1)$$

This entails 4 partial multiplications: $u_1 v_1$, $u_0 v_1$, $u_1 v_0$ and $u_0 v_0$. The factors 10^N and $10^{N/2}$ do not require any multiplication but merely indicate simple leftwards shifts by N and $N/2$ digits respectively.

11.2 Karatsuba multiplication

By comparison, the Karatsuba multiplication constructs the product of u and v as follows:

$$uv = (10^N + 10^{N/2})u_1 v_1 + 10^{N/2}(u_1 - u_0)(v_0 - v_1) + \\ + (10^{N/2} + 1)u_0 v_0 \qquad (11.2)$$

Hey presto! Suddenly only 3 partial multiplications are necessary, namely for the products $u_1 v_1$, $(u_1 - u_0)(v_0 - v_1)$ and $u_0 v_0$. This is a considerable advance, which is only slightly offset by three additional additions.

A numeric example will make this clearer. Let $N = 4$, $u = 9876$, $u_1 = 98$, $u_0 = 76$, $v = 5432$, $v_1 = 54$, $v_0 = 32$..

School Multiplication		Karatsuba Multiplication	
98·54	5292	98·54	5292
76·54	4104	Ditto	5292
98·32	3136	$(98 - 76)(32 - 54)$	-0484
76·32	24 32	76·32	2432
		Ditto	2432
	536464 32		53646432

In the course of this type of multiplication, three multiplications are executed with half-length operands. Each of these contains three further multiplications with quarter-length operands etc. Karatsuba multiplication thus lends itself well to a recursive call. This will be seen from the following verbal piece of code.

Algorithm (11.3) Karatsuba multiplication (alg 11.3)

The product of the numbers in u and v is calculated and stored in r. Variables u and v are of length n, r is of length $2n$. n must be an even number. The upper (left) half of u is called u_1 and the lower (right) half is called u_0. The corresponding two halves of v are v_1 and v_1. A temporary store t of length $2n$ is required.

```
karatsuba(r, u, v, n)
{
    if (n < limit or n is not an even number)
        multiply in the normal way,
    else
    {
        call karatsuba(r, u, v, n/2) to calculate u₁ · v₁
            and store result in the left-hand half of r.
        call karatsuba(r+n, u+n/2, v+n/2, n/2) to
            calculate u₀ · v₀ in the right-hand half of r.
        copy r to t
        add u₁ · v₁ (or t) and u₀ · v₀ (or t + n)
            to r + n/2, i.e. to the middle two quarters of r.
        calculate |u₁ − u₀| in t and |v₀ − v₁| in t + n/2
        note down the signs.
        call karatsuba(t+n, t, t+n/2, n/2) to calculate
            |u₁ − u₀||v₀ − v₁| in t + n.
        if u₁ − u₀ and v₀ − v₁ have the same sign,
            add |u₁ − u₀| · |v₀ − v₁| to r + n/2, otherwise
            subtract it from r + n/2.
    }
    return
}
```

You will notice the three occurrences of the recursive call of the karatsuba() function.

The variable *limit* is at least $= 1$, but in practice it is often greater than this. In fact, if the operands become too small, Karatsuba multiplication ceases to be worthwhile and is then replaced by a normal multiplication operation.

The same applies if the length n is an odd number (either from the start or after the halving steps). "Awkward" operand length are pure poison to the performance of the Karatsuba multiplication method, which only works effectively when number lengths are powers of two. This characteristic it shares with FFT multiplication, which is described in the next section.

Our demo implementation of the method, which you will find in **mult/karamult** on the CD-ROM, works like this as well.

As we have seen, the beneficial effect of reducing the number of partial multiplications from 4 to 3 is propagated in succeeding steps of the procedure. In this way, Karatsuba multiplication requires a total of only 3^k short multiplications for $N = 2^k$-digit operands, instead of the $N^2 = 4^k$ under the school method. As a result, the total time is reduced from a quadratic order N^2 to the order $N^{\log_2 3} = N^{1.58496...}$. Applied to the above example, the multiplication of two numbers both one million digits long no longer takes a day but a mere 5 minutes!

In view of its simplicity, it is surprising that the Karatsuba method of multiplication is only 40 years old. Neither the ancient astronomers nor the excellent mathematicians who lived in the centuries preceding the invention of the computer, nor even the famous "human computers" of the 18th and 19th centuries seem to have had any method more sophisticated than school multiplication at their disposal, despite their staggering mathematical feats. At least, none of these eminent authorities reported otherwise [77, p. 295].

11.3 FFT multiplication

The Karatsuba method may be good, but actually there are faster methods.

The key to greater speed is to perform a *transformation*. Transformations often work wonders, not just in mathematics. If a problem cannot be resolved satisfactorily by a direct route, then, with luck, it will be easier to solve if it is transformed into a different problem. In the present case, this proved to be so.

Prior to the multiplication, the multiplicands are first changed, or rather transformed, in a particular way. An operation which is equivalent to normal multiplication but runs more quickly is then executed on the transformed multiplicands. The results are subsequently transformed back, producing the desired product.

Naturally the detour should not take longer than the direct route.

Perhaps the easiest example of a helpful transformation is one which could have been reported hundreds of years ago. Imagine how laborious it would be for our method to perform even simple multiplications using the Roman numeral system, e.g. to multiply MMMDCCLXXXIX by XLIX. Perhaps one day an Arab looked over the shoulder of a Roman and, surprised to see so much unnecessary work, decided to help him: he *transformed* the two starting Roman numbers into their Arabic equivalents, then multiplied them using the Arabic decimal system, just as we do today, and then transformed the Arabic result back into Roman numerals. Anyone can see that this method would have achieved the desired outcome far more quickly.

No doubt you will remember from your schooldays another multiplication procedure which also works with transformations. This entails using logarithms. Before the pocket calculator won the day, this method was *the* preferred aid to multiplication, especially as it was possible to make things even easier with the famous slide rule.

Under this method, first of all the two multiplicands were *turned into logarithms* using logarithmic tables, then these logarithms were *added*, and finally the sum was *converted back*. Anyone who had to do this will be able to remember his astonishment when the product he was after actually emerged at the end.

Expressed graphically, the logarithmic transformation goes as follows:

$$
\begin{array}{ccc}
a, b & \to \text{Take logarithms} \to & \log a, \log b \\
\downarrow & & \downarrow \\
\text{Multiplication} & & \text{Addition} \\
\downarrow & & \downarrow \\
a \times b & \leftarrow \text{Convert back} \leftarrow & \log a + \log b
\end{array}
\qquad (11.4)
$$

Instead of proceeding along the direct path from a, b to the product $a \times b$, this type of multiplication makes a detour via the logarithms of a and b. Here it is making use of the logarithm theorem, according to which $\log(a \times b) = \log a + \log b$. The logarithms of a and b are therefore added, following which they are converted back to ordinary numbers to give the result.

Instead of proceeding along the direct path from a, b to the product $a \times b$, this type of multiplication makes a detour via the logarithms of a and b. Here it is making use of the logarithm theorem, according to which $\log(a \times b) = \log a + \log b$. The logarithms of a and b are therefore added, following which they are converted back to ordinary numbers to give the result.

Multiplication using logarithms may be elegant for applications in which speed is all-important and storage is limited, however it is not considered, simply because there is a quicker method. This is *multiplication using fast Fourier transforms*, known as *FFT multiplication*. It is the fastest known method of multiplying long numbers and is therefore used in all the big π computations. The procedure is described in detail below.

FFT multiplication also takes a detour involving transformations, as a means of achieving the desired result in less time. The transformation involved is the *Fourier transformation* or *Fourier transform*, as it is generally called, (FT) and its rapid variant is called the *fast Fourier transform* (FFT).

Fourier transforms date back to the French mathematician Jean Baptiste de Fourier (1788–1830) who in his price-winning essay *Théorie analytique de la chaleur* (Analytical Theory of Heat) (1822) proposed them as a means of analysing series. They have since become a common and indispensable tool in the natural sciences.

Fast Fourier transformation is generally dated to the year 1965, in which J. Cooley and J. Tukey [43] presented an FFT algorithm to the astonished expert world for the first time (see page 198). It was subsequently established that such algorithms had been used before and that the roots are even to be found in the work of C.F. Gauss (1805) [65].

The commutative diagram of FFT multiplication looks like this:

$$
\begin{array}{ccc}
a, b & \rightarrow \text{forward FFT} \rightarrow & F(a), F(b) \\
\downarrow & & \downarrow \\
\text{"School"} & & \text{elementwise} \\
\text{multiplication} & & \text{multiplication} \\
\downarrow & & \downarrow \\
a \times b & \leftarrow \text{backward FFT} \leftarrow & F(a) \cdot F(b)
\end{array}
\qquad (11.5)
$$

In the representation shown above, "forward FFT" corresponds to the conversion into logarithms and "backward FFT" cooresponds to the conversion out of logarithms in the previous example (11.4). Both methods also entail three transformations. These are the two forward

FFT's for the two multiplicands and one backward FFT for their product.

The essential ingredients of FFT multiplication are set out below.

Polynomials. A polynomial (of degree N in an Unknown x) can generally be written as $P(x) = \sum_{k=0}^{N} p_k x^k$. The p_k are called the coefficients of the polynomial and are always integers here. Examples of polynomials are $A(x) = 8x^3 + 7x^2 + 8x + 9$ or $B(x) = 4x^2 - x + 2$.

Multiplication of polynomials. We use these examples $A(x)$ and $B(x)$, and multiply them into a product polynomial $C(x)$:

$$
\begin{aligned}
C(x) = \quad & A(x) \quad \times \quad B(x) \\
= \quad & (8x^3 + 7x^2 + 8x + 9) \times \quad (4x^2 - x + 2) \\
\hline
= \quad & 32x^5 + 28x^4 \quad +32x^3 \quad +36x^2 \\
& \quad\quad\quad -8x^4 \quad -7x^3 \quad -8x^2 \quad -9x \\
& \quad\quad\quad\quad\quad\quad\quad +16x^3 \quad +14x^2 \quad +16x \quad +18 \\
\hline
= \quad & 32x^5 + 20x^4 \quad +41x^3 \quad +42x^2 \quad +7x \quad +18
\end{aligned}
$$

The general formula for the coefficients c_k of the polynomial $C(x) = \sum_{k=0}^{2N} c_i x^j$, which is the product of $A(x) = \sum_{i=0}^{N} a_i x^i$ and $B(x) = \sum_{j=0}^{N} b_j x^j$, is quite simple: $c_k = \sum_{i+j=k} a_i b_j$. To convince oneself that this is correct, one needs only to write out the product for the first few powers of x and compare the coefficients. The elegant name for the sequence of c_k's is the *linear convolution* of the sequences of a_i and b_j. This convolution corresponds to the multiplication in the space of the real numbers.

Numbers are (almost) polynomials. A number like 6789 can also be expressed as $6 \cdot 10^3 + 7 \cdot 10^2 + 8 \cdot 10^1 + 9 \cdot 10^0$. Generally the decimal number $d_N d_{N-1} \ldots d_2 d_1 d_0$ is equal to $\sum_{k=0}^{N} d_k 10^k$. The digits of a decimal number are thus the coefficients of a polynomial $P(x)$ which for $x = 10$ has exactly the value of that decimal number. It is because these coefficients are restricted here to the values $0, 1, \ldots 9$, that the word "almost" is included in the section title.

Number multiplication = polynomial multiplication + carryovers. Numbers can be multiplied through multiplication of the corresponding polynomials. However the c_k values cannot be ≥ 10. But since the allowed values of d_k are not restricted to $0 \ldots 9$, after the multiplication carryovers have to be formed from left to right and resolved.

As an example, consider the polynomial multiplication of 67 times 89, during which such carryovers occur in three of the four c_k.

a_i, b_j :		67	×	89
$a_i \times b_1$:		48	56	
$a_i \times b_0$:			54	63
c_k :		24	110	63
d_0 :			6	3
d_1 :		11	6	
d_2 :	3	5		
d_3	:0	3		

Fourier transformation, forward and backwards. The forward Fourier transform $F(a)$ of a digit string $a = \{a_0, a_1, a_2, \ldots, a_{N-1}\}$ is a digit string of equal length with the components $\hat{a}_0, \hat{a}_1, \hat{a}_2, \ldots, \hat{a}_{N-1}$, which are defined as follows:

$$\hat{a}_\omega := \sum_{k=0}^{N-1} a_k(\cos(2\pi\omega k/N) + i\sin(2\pi\omega k/N)) = \sum_{k=0}^{N-1} a_k e^{2\pi i\omega k/N}$$

The backward Fourier transform $F^{-1}(a)$ of a sequence of numbers a is likewise a number sequence of equal length, but with the components $\hat{a}_\omega := \frac{1}{N}\sum_{k=0}^{N-1} a_k e^{-2\pi i\omega k/N}$. The difference here is the initial factor $\frac{1}{N}$ and the minus sign in the exponent of the coefficients.

The direct implementation of these FT definitions leads to the slow Fourier transform. It consists of two simple loops. The calculation is performed with complex numbers.

```
// --------------------------
// sft() (slow) fourier transform
// n      length of array f, not nec. a power of 2
// isign determines forward (> 0) or backward (< 0) transform

void sft(Complex *f, long n, int isign)
{
    Complex *res = new Complex[n];
    const double ph0 = isign > 0 ? +2.0*M_PI/n : -2.0*M_PI/n;

    for (long w=0; w<n; ++w)
    {
        Complex t = 0.0;
        for (long k=0; k<n; ++k)
        {
            double phi = ph0*k*w;
            t +=  f[k] * Complex(cos(phi), sin(phi));
        }
```

```
            res[w] = t;
        }

        for (long k=0; k<n; ++k)   f[k] = res[k];

        // back transform requires normalization by n
        if (isign < 0)
            for (long k=0; k<n; ++k) f[k] /= n;

        delete [] res;
}
```

There are two different reasons why the above abstract definitions of the FT, apparently a matter of luck, are the key to rapid multiplication: firstly the convolution theorem and secondly the existence of fast techniques for the calculation of Fourier transforms.

Convolution theorem. The *convolution theorem* $F(a \times b) = F(a) \cdot F(b)$ applies to Fourier transformed numbers. It states that the customary multiplication of polynomials (i.e. the convolution of the sequences of coefficients with computation time proportional to N^2) is equivalent to an elementwise multiplication of the Fourier transformed coefficient sequences of multiplicands, for which the computation time is only proportional to N. $c_k = \sum_{i+j=k} a_i b_j \Leftrightarrow \hat{c}_y = \hat{a}_y \hat{b}_y$ or, expressed otherwise, $c = F^{-1}(F(a) \cdot F(b))$. Here, c refers to the convolution product (or, to be more accurate, the coefficient sequence of the convolution product) and, by analogy, a denotes the sequence $a_0, a_1, \ldots, a_{N-1}$. The multiplication implied by the dot must be carried out elementwise. The last equation is a different way of writing the same thing which is expressed in the right-hand column of the commutative diagram of the FFT multiplication (11.5). The proof for the convolution theorem is not difficult, and you can find it in the text **fxtrem/fxtrem.*** on our CD-ROM.

Fast Fourier transform (FFT). The time it takes to perform the described (slow) Fourier transform of a digit string of length N is evidently proportional to N^2.

If there were no more efficient method than this, FT multiplication would not get one anywhere. But fortunately the fast Fourier transform (FFT) is available, along with many variants of it. They all have in common that the computational time is reduced from being proportional to N^2 to only $N \log_2(N)$, i.e. only a little worse than proportional to N. (Strictly speaking, the computational effort declines to the order of magnitude of $N log_2(N) log_2(log_2(N))$; this was the result which Schönhage and Strassen revealed in their classic paper on the *Fast multiplication of large numbers* [104] of 1971.) The basic idea is the same as the idea behind the Karatsuba multiplication method: one FT of length N is split into less than four FTs of length $N/2$ (with exactly four FTs there would be no gain). Under the Karatsuba multiplication method, the N-digit multiplication has to be replaced by *three* $N/2$-digit multiplications, under FFT by only *two* FTs of length $N/2$. Here too the splitting step is applied *recursively* until the length reaches one. However an FT of length one is only a "do-nothing": $\hat{a}_0 = \sum_{k=0}^{0} a_k e^{2\pi i \omega k/1} = a_0$. In this way every FT can be calculated through incessant division into shorter sequences and correct assembly of the partial results.

Cyclic and linear convolution. A final ingredient is the conversion of cyclic to linear convolutions. In the formula for the convolution theorem we omitted to mention that the indices k must be understood as cyclic *modulo N*, so it should really be $c_k = \sum_{i+j \equiv k \bmod N} a_i b_j \Leftrightarrow \hat{c}_y = \hat{a}_y \hat{b}_y$. To achieve the required linear convolutions from this cyclic convolution, one lengthens the sequences $a_0, a_1, \ldots, a_{N-1}$ and $b_0, b_1, \ldots, b_{N-1}$ by N zeros to $a_0, a_1, \ldots, a_{N-1}, 0, 0, \ldots, 0$ and $b_0, b_1, \ldots, b_{N-1}, 0, 0, \ldots 0$. It makes no difference here whether the zeros are added on the left or right. The cyclic convolution of *these* sequences produces the linear convolution $c_0, c_1, \ldots, c_{2N-2}, c_{2N-1}$ of the original sequences.

FFT multiplication in action. The demo program shown below constitutes a complete multiplication with the aid of fast Fourier transforms.

```
//
// Fast Fourier Transform (FFT) multiplication in action
//

#include <iostream.h>
#include <Complex.h>

typedef complex<double> Complex;

// function prototypes:

// fft(): fast fourier transform
void fft(Complex f[], long n, int isign);              // see below
// getdigs(): read digits from stdin into array a[]
long getdigs(char *s, long base, long a[], long d_max);  // getdigs.cc
void print(const char *bla, long a[]    , long n);      // print.cc
void print(const char *bla, Complex f[], long n);       // print.cc
```

11. Arithmetic

```cpp
int main()
{
    cout << "\nDEMO OF THE FAST FOURIER TRANSFORM MULTIPLICATION\n";

    typedef unsigned long Digit;
    const long    nab_max = 1000;           // max Digit's of each multplier
    Digit         a[nab_max], b[nab_max];   // Digit's of the multipliers
    Digit         base = 10;
    long          na, nb;                   // length of the multipliers

    while ( 1 )   // main loop
    {
    // 1) Read the two multipliers, set-up length values
        na = getdigs("Multiplicand ", base, a, nab_max);
        if (na == 0) break;
        nb = getdigs("Multiplicator", base, b, nab_max);
        if (nb == 0) break;
        long nc = na + nb;  // length of result array c
        long nf;            // length of complex arrays fa, fb


    // 2) Find nf = smallest power-of-2 >= nc
    //     In case of sft(): letting nf = nc; would suffice.
        for (nf = 2; nf < nc; nf += nf)
            ;
        cout << "Base=" << base
             << ", na=" << na << ", nb=" << nb
             << ", nc=" << nc << ", nf=" << nf << endl;
    // 3) Copy a[] to complex array fa[], also b[] to fb[].
    //     Justify left, pad with zeros to the right:
        Complex *fa = new Complex[nf];
        for (long k=0;  k < na; ++k)   fa[k] = a[k];
        for (long k=na; k < nf; ++k)   fa[k] = 0;
        print("Complex Multiplicand  (zero padded right): ", fa, nf);
        Complex *fb = new Complex[nf];
        for (long k=0 ; k < nb; ++k)   fb[k] = b[k];
        for (long k=nb; k < nf; ++k)   fb[k] = 0;
        print("Complex Multiplicator (zero padded right): ", fb, nf);
    // 4) Perform fast fourier transforms forward on fa[] and fb[]
        fft(fa, nf, +1);
        print("Complex Multiplicand  (after forward FFT): ", fa, nf);
        fft(fb, nf, +1);
        print("Complex Multiplicator (after backward FFT): ", fb, nf);
    // 5) Multiply elementwise: fa[k] = fa[k] * fb[k]
        for (long k=0; k < nf; ++k)    fa[k] *= fb[k];
        print("After elementwise complex multiplication : ", fa, nf);
    // 6) Perform fast fourier transform back incl. normalization:
        fft(fa, nf, -1);
        print("Complex Product (after backwards FFT)    : ", fa, nf);
    // 7) Copy result in fa[] to a real array c[]:
        Digit *c = new Digit[nc];
        for (long k=0; k < nc; ++k)  c[k] = (Digit)(fa[k].real()+0.5);
        print("Real Product                             : ", c, nc);
    // 8) Shift right by one; the least significant digit is c[nc-2]
        for (long k=nc-2; k>=0; --k)   c[k+1] = c[k];
        c[0] = 0;
        print("After right shift by one                 : ", c, nc);
    // 9) carry operation:
        for (long k = nc-1; k >= 1; --k)
```

```cpp
            {
                Digit carry = c[k] / base;
                c[k]       -= carry * base;
                c[k-1]     += carry;
            }
            print("Final Product (after carry operation)   : ", c, nc);
            cout << endl;
            delete [] fa, fb;
            delete [] c;
        }
        return 0;
    }
    //
    // fft() : fast fourier transform
    //         (radix2, decimation in frequency)
    // n       length of array f, must be a power of 2
    // isign   determines forward (> 0) or backward (< 0) transform

    void fft(Complex f[], long n, int isign)
    {
        double pi = isign >= 0 ? +M_PI : -M_PI;

        for (long m=n; m > 1; m /= 2)
        {
            long mh = m / 2;
            double phi = pi / mh;

            for (long j=0; j < mh; ++j)
            {
                double  p  = phi * j;
                Complex cs = Complex(cos(p), sin(p));

                for (long t1=j; t1 < j+n; t1 += m) {
                    Complex u = f[t1];
                    Complex v = f[t1+mh];
                    f[t1]    = u+v;
                    f[t1+mh] = (u-v) * cs;
                }
            }
        }

        // data reordering:
        for (long m=1, j=0; m < n-1; ++m)
        {
            for (long k=n>>1; (!((j^=k)&k)); k>>=1)  {;}
            if ( j > m )  // SWAP(f[m], f[j]);
            {
                Complex t = f[m];
                f[m] = f[j];
                f[j] = t;
            }
        }
        if ( isign < 0 )  // normalise if backwards transform
            for (long k=0; k < n; ++k)  f[k] /= n;
    }
    // --------------------------
```

In the program, the function fft(), which performs the fast Fourier transform, is used instead of the function sft() (slow Fourier transform) shown on page 139. Its source code is shown at the end of the program listing. Compared with the slow variants, in the fast variant the length of the complex arrays must be a power of two (the smallest $\geq na + nb$), because – as mentioned previously – the speed is achieved through recursive halving of the multiplicands. After forward and backward transforms, the results are shifted by one digit towards the right because the transformation leaves the right-most place empty (the convolution product of digits i and j moves to position $i+j$, and simultaneously the convolution product of the right-most places $na-1$ and $nb-1$ move to position $na + nb - 2$). Finally the carryovers are resolved. The other operations in the program should be clear from the foregoing and from the comment statements within the program itself.

Comparison. The fast Fourier transform passes its speed on to the FFT multiplication, in which it is called three times. For this reason, the time taken to perform an FFT multiplication rises only proportionally $N \log_2 N$ to the length N of the multiplicands. The effect is considerable when N is large, as the following graph demonstrates.

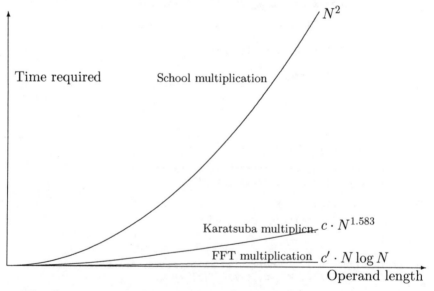

The diagram shows clearly how superior FFT multiplication is to the Karatsuba and school multiplication methods.

In our example of the multiplication of two one-million-digit ($= N$) numbers, the number of individual multiplication operations is reduced from 1 billion to approx. 20 million ($= N \log_2 N$), reducing the time taken to perform the calculation from more than a day to only 3 seconds. It beats the Karatsuba multiplication method, which itself is pretty fast, by a factor of around 100. (You can observe these calculation times for yourself if you run our demo program mult/speedcmp).

11.4 Division

The conventional method of performing long division as taught in schools should not be used to divide long numbers u and v. Instead it is replaced by two operations, namely one multiplication and one reciprocation, in accordance with the following schema:

$$\frac{u}{v} = u \cdot \frac{1}{v} \tag{11.6}$$

We have already dealt in detail with multiplication operations. For the reciprocation $\frac{1}{v}$ of long numbers, the method still used today is that devised more than 300 years ago by Isaac Newton (1642–1727) for determining a zero of a function. As long as certain conditions for continuity are satisfied, then Newton's method determines a zero of a function $f(x)$ iteratively with

$$x_{i+1} = x_i - \frac{f(x_i)}{f'(x_i)} \tag{11.7}$$

Starting from a starting approximation x_0 with $f(x_0) \approx 0$, this rule produces successive values $x_1, x_2, \ldots$, which tend ever closer to zero.

What is unusual about Newton's method is that it converges *quadratically* and on top of that it is *self-correcting*. At each iteration the number of accurate places of the successor value is doubled, and it is sufficient here for the predecessor value to be accurate only to half the number of digits.

In our case, we have to use the function $f(x) = v - 1/x$. Its zero is the desired inverse value $1/v$. With the first derivation $f'(x) = -1/x^2$, the iteration rule runs simply as follows:

$$x_{i+1} = x_i + x_i(1 - v \cdot x_i) \tag{11.8}$$

If the x_i are accurate to m places, then the successor values x_{i+1} will be accurate to at least $2m - 1$ places. Normally one continues the iterations until the expression $x_i(1 - v \cdot x_i) = 0$ within the bounds of the calculation accuracy. Here, the expression (11.8) is calculated economically just as it stands there, and not as $x_{i+1} = x_i(2 - vx_i)$: the term $1 - v \cdot x_i$ is so close to 0 that its left half consists of nothing but zeros, and one therefore only has to multiply with half the digits. An example will illustrate this more clearly.

Suppose we want to calculate $1/v = 1/7 = 0.14285\,714\ldots$ to 8 decimal places. Let the initial value be $x_0 = 0.1$. We then obtain:

k	x_i	$x_i(1 - v \cdot x_i)$	x_{i+1}	Accurate Places
0	0.1	0.03	0.13	1
1	0.13	0.0117	0.1417	2
2	0.1417	0.0011477	0.14284777	4
3	0.14284777	0.00009372	0.14285714	8

8 accurate decimal places are obtained after only three iterations each involving two multiplications of 1, 2 and 4 digits.

To determine the amount of time required, for the sake of simplicity we will assume that the desired accuracy of places n of the last approximation is exactly a power of two 2^k times the accuracy of places n_0 of the starting approximation. In that case, k iterations are needed, and their initial accuracy does not need to be greater than $n_0, 2n_0, 4n_0, \ldots, 2^{k-1}n_0$. With the known summation properties of powers of two $\sum_{i=0}^{k-1} 2^i = 2^k - 1$ and the fact that only multiplications with almost linear time characteristics are used in the iterations steps, in the ideal case (asymptotic) we obtain for the calculation of the inverse value $1/v$ a time requirement of two multiplications of length n times the end accuracy. At the end there is another such long multiplication, so that for the division u/v (asymptotic) 3 long multiplications are necessary.

For various reasons it is not possible to perform the long division in this ideal time when actually implemented. Realistically one would have to estimate the time at four rather than three long multiplications.

11.5 Square root

As one can see from the modern iterative formulae for the π computation, for example the Gauss AGM algorithm, a square root $\sqrt{d}$ with a long radicand d occurs there at least once per iteration, and this has to be calculated for the full length of the results.

If you can still remember how square roots were calculated at school, then forget it as there are some far more effective methods of performing this calculation.

Newton iteration without divisions. This procedure divides the problem into two steps, rather like in the division, the first being the calculation of the reciprocal of the square root you are looking for, $1/\sqrt{d}$, and the second being its multiplication by d, since $d/\sqrt{d} = \sqrt{d}$.

For $1/\sqrt{d}$, it is then necessary to determine the zero of the function $f(x) = 1/x^2 - d$, and for this Newton's method (11.7) provides the following iteration rule:

$$x_{i+1} = x_i + x_i \frac{1 - d \cdot x_i^2}{2} \tag{11.9}$$

The iterations normally end when the expression $x_i(1 - d \cdot x_i^2)/2 = 0$, within the bounds of the calculation accuracy.

x_i converges quadratically, as in the long division. Once again it is possible to reduce the amount of calculation effort by only carrying along the number of digits which are really necessary. Adopting this approach, it takes about the same time to calculate $1/\sqrt{d}$ as it does to perform 3 long multiplications. The calculation of $\sqrt{d}$ itself entails another long multiplication.

Coupled Newton iteration. Arnold Schönhage succeeded in arriving at a significantly more effective solution on the basis of an apparently inferior approach [104, p. 257]. His method calculates the square root $\sqrt{d}$ directly and ascertains the zero of the function $f(x) = d - x^2$, for which from (11.7) the following iteration rule is derived:

$$x_{i+1} = x_i + \frac{d - x_i^2}{2x_i} \tag{11.10}$$

Each step of this contains an expensive division, which "really" disqualifies the procedure. However, there is a wonderful trick which reduces this drawback so effectively that the total time required is less than under the first method we mentioned. This is known as the *coupled Newton iteration* and works with the following double iteration:

Algorithm (11.11) Square root by coupled Newton iteration (alg 11.11)

Initialise:
$$x_0 :\approx \sqrt{d}$$
$$v_0 := 1/(2x_0)$$

Iterate:
$$v_{i+1} := v_i + (1 - 2x_i v_i) \cdot v_i \xrightarrow{2} 1/(2\sqrt{d})$$
$$x_{i+1} := x_i + (d - x_i^2) \cdot v_{i+1} \xrightarrow{2} \sqrt{d}$$

Instead of dividing, the procedure entails multiplying by an approximate reciprocal $v_{i+1} = 1/(2x_i)$ which is improved before the start of the current iteration in a coupled iteration from v_i and x_i.

The time required to perform this calculation is asymptotically 7/3 long multiplications, which in reality expand to 3 long multiplications. For more about this, see the implementation contained in the **hfloat** library under `hfloat/src/hf/itsqrt.cc`

A recursive *MuPAD* procedure for calculating the square root looks like this (after [4]):

```
Sqrt_Coupled_Newton := proc(d, n)
   // assume d >= 0, n >= 1
   local m, g, h, x, e, f, t; // v is static! begin
   t := DIGITS;    // save global precision
   if n <= 2
   then
      DIGITS := n;
      x := sqrt(d); // Initial value of x
      DIGITS := t;
      return(x);
   end_if;
   m := floor(n/2)+1;
   x := Sqrt_Coupled_Newton(d, m);
   DIGITS := m;
   if m <= 2
   then
      v := 1/(2*x);
   else
      g := 1 - 2*x*v;
      h := g * v;   // v from the previous iteration
      v := v + h;   // now v is of precision m
   end_if;
   DIGITS := n;
   e := d - x^2;
   f := e * v;
   x := x + f;
   DIGITS := t;    // restore global precision
   return(x); end_proc:
```

The variable `DIGITS:=` is used to specify the degree of precision required (in decimal places).

11.6 nth root

Some π algorithms also use roots with larger exponents, i.e. expressions of the form $\sqrt[n]{d}$ with $n > 2$. This n is usually a constant and moreover very small, so that one can afford to call a separate algorithm for every n.

There is an interesting procedure for the construction of such algorithms for any n, whose convergence moreover is user-specifiable.

In every case, the procedure entails calculating the reciprocal of the desired nth root at the start - as for the square root. The desired result is then determined via the detour

$$\sqrt[n]{d} = \frac{1}{\sqrt[n]{d^{n-1}}} \cdot d \frac{1}{\sqrt[n]{D}} \cdot d \tag{11.12}$$

and once the root has been calculated, another multiplication by d is necessary at the end.

The starting point for this class of algorithm is the identity

$$\frac{1}{\sqrt[n]{D}} = \frac{x}{\sqrt[n]{1-(1-x^n \cdot D)}}$$
$$= x \cdot (1-y)^{-1/n} \qquad \text{with } y := 1 - x^n D \tag{11.13}$$

Converting to a Taylor series produces

$$\frac{1}{\sqrt[n]{D}} = x \left(1 + \frac{y}{n} + \frac{(1+n)y^2}{2n^2} + \frac{(1+n)(1+2n)y^3}{6n^3} + \ldots + \right.$$
$$\left. + \frac{(1+n)(1+2n)\cdots(1+(k-1)n)y^k}{k!n^k} + \cdots \right) \tag{11.14}$$

The desired iteration rule for a convergence order k is now obtained by terminating this series after the $(k-1)$th term. Let the resulting sum be designated $\Phi_k(n, x)$. The iteration then goes as follows:

$$x_{i+1} = \Phi_k(n, x_i) \tag{11.15}$$

In this way, for example, we obtain an algorithm that converges quadratically for the reciprocal of a square root ($n = 2, D = d$):

$$x_{i+1} = \Phi_2(2, x_i)$$
$$= x_i + x_i \frac{(1 - d \cdot x_i^2)}{2} \tag{11.16}$$

in fact, precisely the iteration rule (11.9) which we saw in the last section.

The following table contains the four iteration rules for the cases $n = 2, 3$ and $k = 2, 3$:

	$\sqrt[2]{\frac{1}{d}}$	$\sqrt[3]{\frac{1}{d}}$
Order of convergence	$D = d$ $y_i = (1 - x_i^2 D)$	$D = d^2$ $y_i = (1 - x_i^3 D)$
2	$x_{i+1} = x_i(1 + y_i/2)$	$x_{i+1} = x_i(1 + y_i/2)$
3	$x_{i+1} = x_i(1 + y_i/2 + 3y_i^2/8)$	$x_{i+1} = x_i(1 + y_i/3 + 3y_i^2/8)$

In this way we have a number of possibilities for calculating the nth roots in an effective manner.

11.7 Series calculation

Calculation of poorly converging infinite series is one of the difficult tasks encountered everyday in working with numbers. The Leibniz series for $\pi/4$ is a particularly glaring case:

$$\frac{\pi}{4} = 1 - \frac{1}{3} + \frac{1}{5} - \frac{1}{7} + -\cdots = \sum_{k=0}^{\infty} (-1)^k \frac{1}{2k+1} \qquad (11.17)$$

This converges so poorly that no less than 10^n terms have to be calculated for n accurate decimal places.

It is all the more surprising that there is an algorithm which speeds up calculation of this series enormously. With this algorithm it is possible, for example, to obtain 1000 accurate decimal places from only 1307 summands, instead of a massive 10^{1000}. Even better, the algorithm can not only be used for the Leibniz series but also for many other series. The only firm condition is that the terms of the series must alternate, i.e. each positive term must be followed by a negative term, which in turn is followed by another positive term. In addition, the series terms have to be simple to calculate and the desired precision of the series sum must not exceed a certain limit, e.g. 1000 decimal places.

This astonishing procedure was developed by Henri Cohen, F. Rodriguez Villegas and Don Zagier [42]. In their paper, the authors describe a whole class of algorithms for other applications. We will confine ourselves here to the most elegant of these. It is aptly named *sumalt*.

11.7 Series calculation

The critical discovery lies in the determination of universal coefficients which are independent of the series to be calculated and by means of which the convergence of the series can be significantly accelerated.

The algorithm approximates the sum $s = \sum_{k=0}^{\infty} a_k$ with alternating a_k through a weighted sum of $a_0, a_1, \ldots, a_{n-1}$ with universal coefficients $c_{n,k}/d_n$. Both $c_{n,k}$ and d_n are integers. For the first n we obtain the following values:

n	d_n	$c_{n,0}$	$c_{n,1}$	$c_{n,2}$	$c_{n,3}$
1	3	$2a_0$			
2	17	$16a_0$	$8a_1$		
3	99	$98a_0$	$80a_1$	$32a_2$	
4	577	$576a_0$	$544a_1$	$384a_2$	$128a_3$

The coefficients do not depend on a_k, but only on the number of terms n, and hence on the degree of precision to which the sum is required to be calculated. For a large class of series a_k the relative error is around $(3+\sqrt{8})^{-n} \approx 5.828^{-n}$, so that for a relative precision of D decimal places the value $n = 1.31 \cdot D$ is sufficient.

The coefficients $c_{n,k}$ are obtained through an iteration rule using an auxiliary variable $b_{n,k}$.

$$b_{n,n-1} := 2^{2n-1} \tag{11.18}$$

$$c_{n,n-1} := b_{n,n-1} \tag{11.19}$$

$$b_{n,k-1} := b_{n,k} \frac{(2k+1)(k+1)}{2(n-k)(n+k)} \quad (k = n-1, n-2, \ldots, 0) \tag{11.20}$$

$$c_{n,k-1} := c_{n,k} + b_{n,k-1} \tag{11.21}$$

We depart from the original article by calculating the coefficients $c_{n,k}$ from $k = n-1$ downwards to $k = -1$, as this makes them easier to calculate and eliminates one square root calculation. Moreover the term d_n, by which the end result has to be divided, is equal to the last calculated c_{n-1}.

In the syntax of the *MuPAD* computer algebra system, the surprisingly simple algorithm runs as follows:

```
sumalt := proc(n)
  local b,c,k,s;
begin
  b := 2^(2*n-1);   // temorary var. b_{n,k}
  c := b;           // c_{n,k}
  s := 0;           //
```

```
for k from n-1 downto 0 do
    t := (-1)^k / (2*k+1);  // here: Leibniz term, general: a_k
    s := s + c * t;
    b := b * ((2*k+1)*(k+1))/(2*(n-k)*(n+k));
    c := c + b;
end_for;
s := s / c;          // last c_{n,k} = d_n
return( s );
end_proc:
```

Just how closely this algorithm approximates the Leibniz series can be seen from the following table:

n	sumalt(n)	Accurate places(D)
1	0.66666 66666 66666 66666 66666 66666	0
2	0.78431 37254 90196 07843 13725 49019	2
5	0.78539 66366 00918 49208 70915 51855	5
10	0.78539 81634 52432 89232 97023 43038	9
20	0.78539 81633 97448 30992 06676 87680	18
30	0.78539 81633 97448 30961 56608 48818	26
$\frac{\pi}{4} =$	0.78539 81633 97448 30961 56608 45819	30

In this way $D = 26$ accurate decimal places of $\pi/4$ were determined with only 30 series terms in the Leibniz series, whereas otherwise this would have required 10^{30} series terms. You can tell just how fast the algorithm is from the fact that it took our PC only 2 seconds to calculate 1000 accurate decimal places.

12. Miscellaneous

12.1 A π quiz

One of the few women who is actively engaged in the "π scene" is Eve A. Andersson. She has put on the Internet a quiz called "The Pi Trivia Game" which gives us the "chance to pay tribute to the magnificent transcendental number that we have all grown to love".[1]

The quiz consists of 25 questions from Eve's "π question database". The whole test is too long to print here, but a few of the questions are provided below:

1. Consider the following series of natural numbers, constructed by successively larger strings of digits from the beginning of the decimal expansion of the number π: 3, 31, 314, 31415, 314159, 3141592, etc. Out of the first 1000 numbers in this series, how many are primes?
   a) 48   b) 34   c) 4   d) 21   e) 58 ?

2. What is another name for π in Germany?
   a) el numero buono
   b) die Ludolphsche Zahl
   c) Gesundheit
   d) die Eulersche Zahl
   e) Drei

3. If one were to find the circumference of a circle the size of the known universe, requiring that the circumference be accurate to within the radius of one proton, how many decimal places of π would need to be used?
   a) 2 million   b) 39   c) 11   d) 48,000   e) 300

---
[1] http://eveander.com/trivia/

4. Among the digits of π currently known, the concentrations of each of the digits 0 - 9 are pretty much equal. However, in the first 30 digits of pi's decimal expansion, one number is conspicuously missing. Which number is it?

a) 7 b) 2 c) 0 d) 8 e) 6

Solutions: 1.: c), 2.: b), 3.: b), 4.: c).

Most people are surprised at the answer to question 1. Among the first 1000 initial sequences of π there are indeed only four prime numbers, and they occur right at the beginning: 3, 31, 314 159 and 3 141 592 653 589 793 238 462 643 383 279 502 884 1. Whether there are any other prime numbers greater than 38 digits long at the beginning of π is not known; at any rate it is certain that there are no other prime numbers less than 5000 digits long.

12.2 Let numbers speak

"If you stare at a number long enough, the number will talk back to you." This conviction of Dario Castellanos is stated in his article *The Ubiquitous π* [41] dated 1988, a collection of "numerological" findings which he and others have made over time.

1. The fraction $\frac{355}{113}$ is known to be a good approximation (to 6 decimal places) of π. The following approximation for $\sqrt{\pi}$

$$\sqrt{\pi} \approx \frac{553}{311 + 1} \tag{12.1}$$

which is almost equal to the fraction $\frac{355}{113}$ read backwards, is less well-known.

2. The following approximation of $\sqrt{\pi}$ contains a rounded version of the first 6 digits of π (3.141593):

$$\sqrt{\pi} \approx \left(\frac{3}{14}\right)^2 \frac{193}{5} \tag{12.2}$$

3. T.E. Lobeck discovered a magical property of π. The square on the left hand side is a standard magic square of the order 5×5. Its lines, columns and diagonals all add up to 65. If one replaces each value n in the square by the nth digit of π, then one obtains a new number square. In this square, all the line totals also occur as column totals:

17	24	1	8	15	2	4	3	6	9	(24)
23	5	7	14	16	6	5	2	7	3	(23)
4	6	13	20	22	1	9	9	4	2	(25)
10	12	19	21	3	3	8	8	6	4	(29)
11	18	25	2	9	5	3	3	1	5	(17)

(17) (29) (25) (24) (23)

Magic square   Square modified with π

12.3 A proof that $\pi = 2$

Look at this sequence of semicircles:

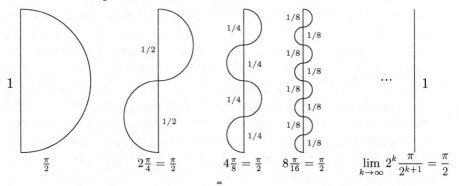

The semi-circle on the left has diameter 1 and the length of its arc is $\frac{\pi}{2}$. The diameters of the two semicircles immediately to the right are half as long, but their arcs are $\frac{\pi}{2}$ long. Again, the arcs of the 4 semicircles in the middle and the 8 semicircles in the last but one diagram also total $\frac{\pi}{2}$ in length. If one continues this an infinite number of times, the wavy line is eventually transformed into a straight line of total length $\frac{\pi}{2}$ which is equal to the total of all the diameters, i.e. 1. Hence,

$$\frac{\pi}{2} = 1$$
$$\pi = 2 \quad \text{q.e.d.}$$

12.4 The big change

From a FORTRAN manual issued by the Xerox company: "The primary purpose of the DATA statement is to specify names for constants;

instead of writing 3.141592653589793 at every place in the program in which π occurs, one can assign this value to a variable PI with a DATA statement and then use that instead of the long form. This also simplifies amending the program *in the event that the value of pi should change.*"

12.5 Almost but not quite

The most well-known series for π is called the Leibniz series and goes as follows:
$$\pi = 4\left(1 - \frac{1}{3} + \frac{1}{5} - \frac{1}{7} + \cdots\right) \tag{12.3}$$
This series is also notorious for its poor convergence which makes it unsuitable for π calculations. After the nth term the error is still $\frac{1}{n}$. To obtain 100 accurate decimal places of π, for example, $\frac{1}{n}$ would have to be equal to 10^{-100}, i.e. $n = 10^{100}$, and that is greater than the number of particles in the entire universe.

However, the series compensates to some extent for its poor convergence with an unusual property. Sometimes after running past the first incorrect digit it continues correctly for a good-sized section. Look at the value which the series produces after 50,000 terms and compare it with the first correct decimal places of π:

$$4 \sum_{k=1}^{50000} \frac{(-1)^{k-1}}{2k-1} = 3.14157\,26535\,89795\,23846\,26423\,83279\,50\ldots \tag{12.4}$$
$$\text{whereas } \pi = 3.14159\,26535\,89793\,23846\,26433\,83279\,50\ldots$$

In the results, the 5th decimal place is already incorrect, and this is not expected to be otherwise in the 50,000th term of the above estimate. But after this incorrect digit, the next 9 digits are entirely correct! Then we get another 8 digits correct, and after another incorrect digit another 8 correct ones. Altogether, only 3 out of the first 33 digits are incorrect (the ones marked as such above).

R.D. North of Colorado Springs, USA, drew this peculiarity to the attention of the Borwein brothers. Martin R. Powell had also noticed this strange feature back in 1982 after 13.5 hours of computer analysis on his school computer (see [93]). The Borweins and their colleague K. Dilcher looked into the phenomenon and clarified the mathematical background in an article [35].

The phenomenon is due to the particular form which the residual sum of the Leibniz series takes, i.e. the sum of all the terms which follow after the $N/2$th series term. This asymptotic behaviour amounts to the following [35]:

$$\pi - 4\sum_{k=1}^{N/2} \frac{(-1)^{k-1}}{2k-1} = \frac{2}{N} - \frac{2}{N^3} + \frac{10}{N^5} - \frac{122}{N^7} + \frac{2770}{N^9} - \frac{101042}{N^{11}} +$$

$$+ \cdots + (-1)^m \frac{2 \cdot E_{2m}}{N^{2m+1}} + \cdots \quad (12.5)$$

J. and P. Borwein, Dilcher,1989

Here the numerators 2, 2, 10, 122, 2770, 101042, ... represent the *Eulerian numbers* E_{2m}, multiplied by 2. N must be divisible by 4.

If we set the value 10^5 for N in (12.5), the first four terms of the residual sum are then $+2 \cdot 10^{-5} = 0.00002$, $-2 \cdot 10^{-15}$, $+1 \cdot 10^{-24}$ and $-1.22\ldots \cdot 10^{-33}$. However, all this means is that the 5th and 15th decimal places are incorrect by -2 and the 24th decimal place is incorrect by $+1$, but the rest of the first 33 places are correct.

In the meantime, the subject has been discussed further in the literature. In [5] we learn that the asymptotic formula (12.5) could have been discovered in only one second if the *Maple* computer algebra system had been fed this line of code:

```
asympt(simplify((sum(-4*(-1)^j/(2*j-1),
                j=1..n/2)-Pi) /(-1)^(n/2+1)),n,8);
```

Before this discovery was made, the Borwein, Borwein and Dilcher article ranked as an example of the superiority of man over computer, but this new finding suggests otherwise.

But, to be fair, one must say that the curiosity discovered by North for decimal places only applies if N is a power of 10. If, for example, only one single series term is added in (12.4), then the result looks quite different:

$$4\sum_{k=1}^{50001} \frac{(-1)^{k-1}}{2k-1} = 3.14161\,26531\,89799\,23842\,26427\,83275\,50\ldots \quad (12.6)$$

Now 8 out of the first 33 decimal places are different from π.

12.6 Why always more?

Why are people always looking for more digits rather than fewer? No doubt this was uppermost in the minds of Wei Gong-yi, Yang Ziquiang, Sun Jia-chang and Li Jia-kai of the Academia Sinica Computer Centre in Peking, who in 1996 published an academic article entitled, *The Computation of π to 10,000,000 digits* [60]. The article made clear the trouble which the four authors had had with their calculation and how they had finally overcome the problems. The article naturally contains a list of references. One of these is another article which sounds familiar, *The Computation of π to 29,360,000 Decimal Digits* by David Bailey, 1988.

It would seem that the direction of progress is not always forwards.

12.7 π and hyperspheres

A circle has the area πr^2 and a sphere the volume $\frac{4}{3}\pi r^3$. In both formulae π appears to the first power.

When considering all multi-dimensional objects, it simplifies matters if one can speak generally in terms of "sphere" and "volume". Thus, for example, a circle may be understood as a 2-dimensional sphere and its area as 2-dimensional volume. Thus, the 1-dimensional circle diameter also has a volume which amounts to $V_1 = 2r$; π does not occur at all in this expression or only to the 0th power.

How does one calculate the volumes of spheres which have more than three dimensions ("hyperspheres")? They can be determined through integral calculus, e.g. using the recursion formula [73, p. 179ff]

$$V_k = V_k(r) = 2\int_0^r V_{k-1}(y)dx \text{ with } y = \sqrt{r^2 - x^2} \quad (12.7)$$

This formula takes us from V_1 to V_2, from V_2 to V_3 and from there in turn to the volumes of higher spheres which are too complex for us to imagine:

$$V_2 = 2\int_0^r V_1(y)dx = 2\int_0^r 2y\,dx = 4\int_0^r \sqrt{r^2 - x^2}\,dx = \pi r^2 \quad (12.8)$$

$$V_3 = 2\int_0^r V_2(y)dx = 2\int_0^r \pi y^2 dx = 2\int_0^r \pi(r^2 - x^2)dx$$

$$= 2\pi(r^2 x - \frac{x^3}{3})\Big|_0^r = \frac{4}{3}\pi r^3 \quad (12.9)$$

12.7 π and hyperspheres

$$V_4 = 2\int_0^r V_3(y)dx = 2\int_0^r \frac{4}{3}\pi y^3 dx = \frac{8}{3}\pi \int_0^r \sqrt{(r^2-x^2)^3}dx$$

$$= \frac{8}{3}\pi(\int_0^r r^2\sqrt{r^2-x^2}dx - \int_0^r x^2\sqrt{r^2-x^2}dx)$$

$$= \frac{8}{3}\pi(\pi\frac{r^4}{4} - \frac{r^4}{4}\arcsin 1) = \frac{8}{3}\pi(\pi\frac{r^4}{4} - \pi\frac{r^4}{16}) = \frac{1}{2}\pi^2 r^4 \qquad (12.10)$$

Interestingly, it is only in the volume of the 4-dimensional sphere that our π appears for the first time *as a square*.

What happens after this? V_5 produces $\frac{8}{15}\pi^2 r^5$, once again with π to the second power. But in $V_6 = \frac{1}{6}\pi^3 r^6$, π then appears to the third power.

The general formulae for the volume and the area of a k-dimensional sphere of radius r go as follows

$$V_k(r) = \frac{1}{(k/2)!}\pi^{k/2}r^k \qquad (12.11)$$

$$O_k(r) = \frac{n}{(k/2)!}\pi^{k/2}r^{k-1} \qquad (12.12)$$

whereby $n!$ is defined recursively:

$$n! = n\cdot(n-1)! \text{ with } 1! = 1 \text{ and } (\tfrac{1}{2})! = \tfrac{1}{2}\sqrt{\pi} \qquad (12.13)$$

It will now be clear how the exponents of π are determined, i.e. as 'half dimension', if necessary rounded to the next lower integer.

This is confirmed in the following list of volumes and areas of selected hyperspheres together with the numeric values in the case $r = 1$:

Dimensions k	Volume $V_k(r)$	$V_k(1)$	Surface $O_k(r)$	$O_k(1)$
1	$2r$	2.0	2	2.0
2	πr^2	3.14	$\pi r/2$	6.28
3	$\pi r^3 \cdot 4/3$	4.19	$\pi r^2 \cdot 4$	12.57
4	$\pi^2 r^4/2$	4.93	$\pi^2 r^3 \cdot 2$	19.74
5	$\pi^2 r^5 \cdot 8/15$	5.26	$\pi^2 r^4 \cdot 8/3$	26.32
6	$\pi^3 r^6/6$	5.17	$\pi^3 r^5$	31.01
7	$\pi^3 r^7 \cdot 16/105$	4.72	$\pi^3 r^6 \cdot 16/15$	33.07
8	$\pi^4 r^8/24$	4.06	$\pi^4 r^7/3$	32.47
9	$\pi^4 r^9 \cdot 32/945$	3.30	$\pi^4 r^8 \cdot 32/105$	29.69
10	$\pi^5 r^{10}/120$	2.55	$\pi^5 r^9/12$	25.50
20	$\pi^{10}r^{20}/10!$	0.0258	$\pi^{10}r^{19}/10!$	0.516
100	$\pi^{50}r^{100}/50!$	$2\cdot 10^{-40}$	$\pi^{50}r^{99}/50!$	$2\cdot 10^{-38}$

In the columns with volumes $V_k(1)$ and surfaces $O_k(1)$ for the unit spheres, it is striking that with 5 and 7 dimensions a maximum occurs up to which the values continue to get bigger but after which they become smaller and converge towards 0. The unit hyperspheres seem to expand initially as the number of dimensions is increased, and thereafter to contract in on themselves, even though their radiuses remain unchanged (= 1). "Imagine that if you can".

Individual sums of these terms are also interesting if one continues them infinitely. Thus, for example, the volumes of the unit hyperspheres with even dimensions add up to a particularly simple expression,

$$\sum_{k=1}^{\infty} V_{2k}(1) = e^{\pi} \tag{12.14}$$

and the total values of the volumes and areas are very nearly integers.

$$\sum_{k=1}^{\infty} V_k(1) = e^{\pi}(1 + \operatorname{erf}(\sqrt{\pi})) = 45.999326\ldots \tag{12.15}$$

$$\sum_{k=1}^{\infty} O_k(1) = 2(1 + \pi e^{\pi}(1 + \operatorname{erf}(\sqrt{\pi}))) = 291.0222\ldots \tag{12.16}$$

where $\operatorname{erf}(\sqrt{\pi}) = 2\sum_{k=0}^{\infty} \frac{(-1)^k \pi^k}{k!(2k+1)}$.

Is that an accident?

One might surmise from the formulae for the 3-dimensional sphere ($\frac{4}{3}r^3\pi$) and other 3-dimensional bodies such as the circular cylinder ($r^2 H \pi$) that π generally occurs to the first power in the volume of smooth 3-dimensional bodies. But in fact this is not the case. For example the torus ("annulus") has the volume $2Rr^2\pi^2$, and here π occurs to the second power.

12.8 Viète × Wallis = Osler

The history of π contains the interesting detail that virtually the two first infinite expressions for π were infinite *products*, whereas almost all the other similar expressions are infinite *sums*. These are the famous products of François Viète (1540–1603), 1593,

$$\frac{2}{\pi} = \sqrt{\frac{1}{2}} \cdot \sqrt{\frac{1}{2} + \frac{1}{2}\sqrt{\frac{1}{2}}} \cdot \sqrt{\frac{1}{2} + \frac{1}{2}\sqrt{\frac{1}{2} + \frac{1}{2}\sqrt{\frac{1}{2}}}} \cdots \quad (12.17)$$

Viète, 1593

and John Wallis (1616–1703), 1655,

$$\frac{2}{\pi} = \frac{1 \cdot 3}{2 \cdot 2} \cdot \frac{3 \cdot 5}{4 \cdot 4} \cdot \frac{5 \cdot 7}{6 \cdot 6} \cdot \frac{7 \cdot 9}{8 \cdot 8} \cdots \quad (12.18)$$

Wallis, 1655

Although these products are several centuries old (the originals looked slightly different), the discovery that they are related to each other and can be united into a single expression is only quite recent. Thomas Osler demonstrated this in October 1999 with the following formula [88]:

$$\frac{2}{\pi} = \prod_{n=1}^{p} \sqrt{\frac{1}{2} + \frac{1}{2}\sqrt{\frac{1}{2} + \frac{1}{2}\sqrt{\frac{1}{2} + \cdots + \frac{1}{2}\sqrt{\frac{1}{2}}}}} \times$$

(n radicals)

$$\times \prod_{n=1}^{\infty} \frac{2^{p+1}n - 1}{2^{p+1}n} \cdot \frac{2^{p+1}n + 1}{2^{p+1}n} \quad (12.19)$$

Osler[88], 1999

Here, each of the two historical products is a special case of a more general double product. The first part of (12.19) consists of the first p factors of the Viète product (12.17), whereas the second part is a "Wallis-like" product. Osler describes it in this way because in the case of $p = 0$ it is exactly equal to the Wallis product (12.18) and in other cases of p it is very similar to the Wallis product, only missing a few partial products.

The Osler product turns into the Viète product as p tends to infinity and is equal to the Wallis product when $p = 0$. In the intermediate cases $p = 1$, $p = 2$ etc., we obtain *combined Viète- and Wallis-like products*:

$$p = 0 : \frac{2}{\pi} = \frac{1 \cdot 3}{2 \cdot 2} \cdot \frac{3 \cdot 5}{4 \cdot 4} \cdot \frac{5 \cdot 7}{6 \cdot 6} \cdot \frac{7 \cdot 9}{8 \cdot 8} \cdot \frac{9 \cdot 11}{10 \cdot 10} \cdot \frac{11 \cdot 13}{12 \cdot 12} \cdots$$

Wallis's original product

$$p = 1 : \frac{2}{\pi} = \sqrt{\frac{1}{2}} \times \frac{3 \cdot 5}{4 \cdot 4} \cdot \frac{7 \cdot 9}{8 \cdot 8} \cdot \frac{11 \cdot 13}{12 \cdot 12} \cdot \frac{15 \cdot 17}{16 \cdot 16} \cdot \frac{19 \cdot 21}{20 \cdot 20} \cdots$$

$$p = 2 : \frac{2}{\pi} = \sqrt{\frac{1}{2}} \cdot \sqrt{\frac{1}{2} + \frac{1}{2}\sqrt{\frac{1}{2}}} \times \frac{7 \cdot 9}{8 \cdot 8} \cdot \frac{15 \cdot 17}{16 \cdot 16} \cdot \frac{23 \cdot 25}{24 \cdot 24} \cdot \frac{31 \cdot 33}{32 \cdot 32} \cdots$$

$$p = 3 : \frac{2}{\pi} = \sqrt{\frac{1}{2}} \cdot \sqrt{\frac{1}{2} + \frac{1}{2}\sqrt{\frac{1}{2}}} \cdot \sqrt{\frac{1}{2} + \frac{1}{2}\sqrt{\frac{1}{2} + \frac{1}{2}\sqrt{\frac{1}{2}}}} \times$$
$$\times \frac{15 \cdot 17}{16 \cdot 16} \cdot \frac{31 \cdot 33}{32 \cdot 32} \cdot \frac{47 \cdot 49}{48 \cdot 48} \cdot \frac{63 \cdot 65}{64 \cdot 64} \cdots$$

$$\cdots$$

$$p \to \infty : \frac{2}{\pi} = \sqrt{\frac{1}{2}} \cdot \sqrt{\frac{1}{2} + \frac{1}{2}\sqrt{\frac{1}{2}}} \cdot \sqrt{\frac{1}{2} + \frac{1}{2}\sqrt{\frac{1}{2} + \frac{1}{2}\sqrt{\frac{1}{2}}}} \cdots$$

Viète's original product

Every time p is increased by 1, a square root is added to the Viète-like part and every second factor in the Wallis-like part is removed.

The Wallis-like part in the Osler products can be interpreted as an error factor of the Viète-like part. One can then see just how accurate the Viète product is when it is terminated after p factors, as the error factor is only around $1 + 4^{-p-1}$, so that, for example, after 25 factors we already have 15 accurate decimal places of π.

To derive his formula, Osler used the same half-angle functions for $\sin x = 2 \sin \frac{x}{2} \cos \frac{x}{2}$ and $\cos \frac{x}{2} = \sqrt{\frac{1}{2}(1 + \cos x)}$ which Viète also used (see page 187) together with the infinite product for $\sin x = x \prod_{n=1}^{\infty} (1 - \frac{x^2}{n^2 \pi^2})$, which Euler discovered some 150 years after Viète.

12.9 Squaring the circle with holes

Let us stay with the two famous π products and now consider a beautiful article by Hansklaus Rummler [99]:

For the individual factors of the Viète product

$$\frac{2}{\pi} = \sqrt{\frac{1}{2}} \cdot \sqrt{\frac{1}{2} + \frac{1}{2}\sqrt{\frac{1}{2}}} \cdot \sqrt{\frac{1}{2} + \frac{1}{2}\sqrt{\frac{1}{2} + \frac{1}{2}\sqrt{\frac{1}{2}}}} \cdots \qquad (12.20)$$

Viète, 1593

there is a simple geometric explanation. If U_n is the perimeter of a regular 2^n-sided polygon which is inscribed in a circle of radius 1, then these factors are equal to the ratio of 2 successive perimeters $U_n : U_{n+1}$. We can now express the right-hand side of (12.20) as follows:

$$\frac{U_1}{U_2} \cdot \frac{U_2}{U_3} \cdot \frac{U_3}{U_4} \cdot \frac{U_4}{U_5} \cdots = \frac{U_1}{U_\infty} \quad (12.21)$$

The Viète formula now becomes immediately clear. The inner numerators and denominators cancel out in pairs, leaving the first numerator U_1 and the last denominator U_∞. U_1 is equal to the perimeter of a digon, i.e. it is equal to twice the diameter 4. U_∞ is the perimeter of a 2^∞-sided polygon and therefore identical to the circumference 2π, hence

$$\frac{U_1}{U_\infty} = \frac{4}{2\pi} = \frac{2}{\pi} \quad (12.22)$$

i.e. equal to the left-hand side of (12.20).

For the Wallis product,

$$\frac{\pi}{4} = \frac{2 \cdot 4}{3 \cdot 3} \cdot \frac{4 \cdot 6}{5 \cdot 5} \cdot \frac{6 \cdot 8}{7 \cdot 7} \cdot \frac{8 \cdot 10}{9 \cdot 9} \cdots \quad (12.23)$$

$$= \prod_{n=1}^{\infty} \left(1 - \frac{1}{(2n+1)^2}\right) \quad (12.24)$$

Wallis, 1655

on the other hand, there is apparently no such geometric proof. But one can interpret the product via a geometric construction, as follows.

Divide a square of side length 1 into 3×3 sub-squares and punch out the innermost square (left-hand picture):

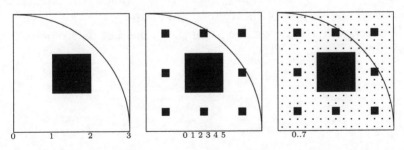

The area of the figure on the left without the black square clearly comes to $1 - \frac{1}{9} = 0.888\ldots$. Now take the 8 remaining sub-squares, divide each one into 5×5 squares and once again punch out the middle of each one. After performing this operation, the total area of the figure is only $(1-\frac{1}{9})\cdot(1-\frac{1}{25}) = 0.853\ldots$ (centre diagram). In the next step, each of the remaining 24 squares is divided into 7×7 squares and once again the centre square of each one is punched out (right-hand diagram).

The process only needs to be continued to infinity to obtain a total area of $\pi/4 = 0.78539816\ldots$. But significantly earlier than this, after perhaps 3 or 4 iterations, one obtains an attractive "Wallis pattern".

12.10 An (in)finite funnel

The curve for the equation $y = 1/x$ is the familiar hyperbola. Consider the right-hand part, from $x = 1$ to ∞. If this section of the curve is rotated about the x-axis, the result is a funnel-shaped object whose *volume* comes to exactly[2] π.

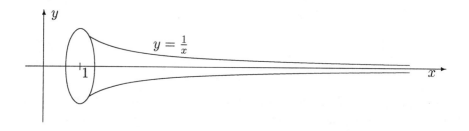

This may not be particularly remarkable in itself. But how about this, then? the *surface* of this funnel[3] is infinite!

In other words, it is possible to fill the funnel but not to paint it in its entirety.

---

[2] $V = \int_1^\infty \frac{\pi}{x^2} dx = -\frac{\pi}{x}\Big|_1^\infty = 0 - -\pi = \pi$

[3] $O = \int_1^\infty \frac{2\pi}{x} dx = 2\pi \ln x\Big|_1^\infty = \infty - 0$

13. The History of π

Mathematical analysis of the circle is one of the oldest challenges to have faced mathematicians. Statements as to how the circumference or the area of a circle can be expressed through other variables are already found in the oldest mathematical documents.

In the beginning, $\pi = 3$. This we may assume without reservations since it corresponds to what one can see by eye. For centuries this value was sufficient for all practical applications, for example in surveying, astronomy and architecture. The real history of π only begins when better approximations than 3 were discovered. This moment in time varies from culture to culture, but by the start of the second millennium BC such approximations had been developed in several places. Since then 4000 years of research into π have taken place, and in fact no single mathematical subject can trace its history so far back into the past.

The symbol π

π research may date back to ancient times, but the name "π" is of relatively recent provenance. It was only in the 18th century AD that the concept came to be referred to with the Greek letter π. Before that it was necessary to describe it in a roundabout way, for example using a description like [114, p. 278] *quantitas, in quam cum multiplicetur dyameter, proveniet circumferentia* (the quantity which, when the diameter is multiplied by it, yields the circumference).

The person who invented the symbol π is assumed to have been the Englishman William Jones (1675–1749), because he used π in the sense as it is used today in his *Synopsis Palmariarum Mathesos*, a compilation of notable mathematical exercises published in 1706. On page 263 he gives the following statement [68]:

...in the *Circle*, the *Diameter* is to *Circumference* as 1 to
$$\frac{16}{3} - \frac{4}{239} - \frac{1}{3}\frac{16}{5^3} - \frac{4}{239^3} + \frac{1}{5}\frac{16}{5^5} - \frac{4}{239^5} - \ldots = 3.14159\ldots = \pi$$

However, neither this nor the other derivations given at this point in the text are actually the work of Jones, but rather they flowed from "the accurate and Ready Pen of the Truly Ingenious Mr. John Machin" (1680–1752), to quote Jones's own Introduction. The symbol π must therefore be attributed to the same man who had already earned his place in history with his arctan formula (5.13) and his record calculation of π to 100 decimal places (see page 192).

Some authors had used the letter π even earlier for other variables relating to the circle. Thus already in 1631 we find it used by William Oughtred (1574–1660), who was one of John Wallis's teachers. In his *Clavis Mathematicae* Oughtred used the symbol π to denote a length equal to half the circumference of a circle and the symbol δ for the half a diameter when he established the proportion $7 : 22 = \delta : \pi = 113 : 355$ [114, p. 302], so that "our" π was the same as "his" $\pi : \delta$. π was also used in a similar way in 1669 by Isaac Barrow (1630–1678)[41, p. 91].

For 30 years Jones's or, more likely, Machin's invention was not repeated by anyone else. Other symbols were tried instead. Johann Bernoulli (1667–1748), for example, employed the Latin letter c and Leonhard Euler (1707–1783) used first p in 1734 and then c in 1736.

A revival of the symbol π occurred when Euler used it again in the treatise he wrote in 1736 entitled *Mechanica sive motus scientia analytice exposita*. From Euler, "not least due to his extensive correspondence" (Tropfke) with all and sundry, the symbol soon spread to other mathematicians as well [114, p. 303]. It had finally won the day when Euler used it in his famous 1748 publication in Latin, *Introductio in analysin infinitorum* (Introduction to the Analysis of the Infinite) [52]. In this work, Euler introduced it with the words, "*For the sake of brevity* we will write this number as π; thus π is equal to half the circumference of a circle of radius 1 or equal to the length of an arc of 180 degrees." [53, p. 95].

Again, the handy and commonly used term "radius", also appeared only relatively recently. Contrary to what one would expect, it dates not from antiquity but first appears in a book published in 1583 [114, p. 134]. However, from that date on it was virtually the only word used for the concept. The Greeks too did not call the radius half-diameter, semi-diameter or another similar term, but instead they referred to "the straight line from the centre point", and this roundabout description continued to be used long into the Middle Ages by faithful imitators of the Greek elements.

13.1 Antiquity

Babylon

A clay tablet unearthed in 1936 from the Old Babylonian period, approx. 1900-1600 BC, states that the circumference of an hexagon is 0;57,36 (in base 60) $= 96/100 = 24/25$ times the circumference of the circumscribed circle [72, p. 18]. From $u_{hexagon} = 3 \cdot d = 24/25 \cdot u_{circle} = 24/25 \cdot \pi \cdot d$ we get what is perhaps the oldest approximation to π,

$$\pi_{Babylon} = 3\frac{1}{8} = 3.125 = \pi - 0.0165\ldots \tag{13.1}$$

Egypt

In ancient times it was not absolutely essential to calculate the circumference of a circle, as it could always be measured. On the other hand it is quite difficult to arrive at an accurate figure for the area of the circle by measurement, so that it was highly desirable to be able to calculate it. Problem no. 50 in the Egyptian *Rhind Papyrus* (named after A. H. Rhind, a Scot who purchased it in 1858 in Luxor), which dates back to around 1850 BC, reads as follows [72, p. 18]: "Example of a round field of diameter 9. What is the area? Take away 1/9 of the diameter; the remainder is 8. Multiply 8 times 8; it makes 64. Therefore, the area is 64." The Egyptian scholar was thus using the formula $A = (d - d/9)^2 = (8d/9)^2$. When this is compared with the formula for the area, $A = \pi \cdot d^2/4$, we obtain the approximation

$$\pi_{Egypt} = \left(\frac{16}{9}\right)^2 = 3.16049\ldots = \pi + 0.0189\ldots \tag{13.2}$$

How this value came about is suggested by problem no. 48 of the Rhind Papyrus in which is shown the figure of an octagon inscribed in a square of side 9. The author may perhaps have observed that the area of the octagon which is $81-4\cdot 3^2/2 = 63$ units was somewhat smaller than the area of the circle, and therefore added on a small amount to correct it [72, p. 19].

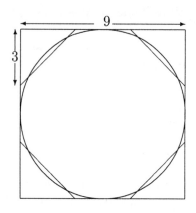

It is possible that values for π either as accurate or even more accurate were known already several centuries earlier in Egypt. This is suggested by evidence that as early as 2600 BC aids existed for the calculation using inscribed irregular polygons. It could be that these were used to derive good approximations for π. For further information on this point see page 209.

India

The Indian *Sulvasutras* (which means "cord rules", i.e. rules for building altars of certain shapes with the aid of pieces of cord [51, p. 102]), which is certainly older than the surviving version dating from 600 BC [72, p. 4], also describes a calculation of the area of a circle: "If you wish to turn a circle into a square, divide the diameter into 8 parts, and again one of these eight parts into 29 parts; of these 29 parts remove 28, and moreover the sixth part (of the one left) less the eighth part (of the sixth part)." This produces a side length s of the required square of

$$s = d\frac{1}{8}\left(7 + \frac{1}{29}\left(1 - \frac{1}{6}\left(1 - \frac{1}{8}\right)\right)\right)$$
$$s = d\frac{9785}{11136}$$

and hence

$$\pi_{India} = 4s^2/d^2 = \left(\frac{9785}{5568}\right)^2 = 3.08832\ldots = \pi - 0.0532\ldots \quad (13.3)$$

Another famous approximation to π which has been ascribed to the Chinese may actually also be of Indian origin. This is

$$\pi \approx \sqrt{10} = 3.16227\ldots \tag{13.4}$$

Tropfke cites a source according to which $\sqrt{10}$ is already found in the year 150 BC in the writings of the Indian scholar Umasvati. The author apparently even believes that he can date this value back to 500 BC in India [114, p. 277].

Bible

The Bible too mentions a value which is an approximation for π, as indeed one would expect. The architect Hiram of Tyre was commissioned by King Solomon built a circular water reservoir out of ore, described as a "molten sea". In 1 Kings 7:23 and again using the same words in 2 Chronicles 4:2, we are told: "And he made a molten sea of ten cubits from brim to brim..., and a line of 30 cubits did compass it round about."

Thirty cubits round about and 10 cubits wide produces a $\pi_{Bible} = 3$. This is pretty pathetic, not only when considered in absolute terms, but also for the time, 550 BC, as more accurate values of π had been known much earlier. Many people have ridiculed the Bible for failing to come up with anything better, inspiring exegetes to search long and hard for statements which would present the Bible in a better light. For example in the 18th century it was declared that the molten sea must have been a hexagon [114, p. 261]. Whether more recent endeavours are any better is somewhat unlikely. One example here is M.D. Stern [111], who concludes from the difference between the spoken and written versions of the Bible that the actual value of π is obtained through division of both values, giving

$$\pi_{Bible} = \frac{333}{106} = 3.141509\ldots = \pi - 0.000083\ldots \tag{13.5}$$

This value is accurate to four decimal places, and if it could only be confirmed, it would certainly silence the mirth at the apparent inaccuracy of the Bible.

Ancient Greece

All the great Greek mathematicians of the classical era from the 5th to the 3rd centuries BC worked on problems relating to circles. One discovery that was especially important for π mathematicians was that of the method of exhaustion. Under this method, which is attributed to Antiphon (c. 430 BC) or Eudoxos (408–355 BC), the area of a two-dimensional figure such as a circle can be arrived at through mental "exhaustion" using ever more elaborate versions of figures whose areas are known, such as polygons. Euclid (c. 330–275 BC) used this method, for example, to prove the theorem that the areas of circles are proportional to the squares of their diameters.

Yet none of the eminent Greek mathematicians, from Anaxagoras (c. 499–428 BC) to Euclid, seems to have improved on the numeric value of π. The Greeks were geometricians; algebra did not interest them. They were more interested in geometric aspects of circles such as the problem of *squaring the circle*. This refers to the problem of how to *construct* a square whose area is equal to the area of a given circle. Several mathematicians had already produced solutions to this problem when Euclid increased the difficulty of the requirement by specifying that it must be solved only with an (unmarked) compass and (unmarked) straight edge. As was only established in 1882, these additional conditions make the squaring the circle insoluble.

Nevertheless the philosopher Plato (427-348 BC) is supposed to have obtained what for his day was a very accurate value for π, $\sqrt{2} + \sqrt{3} = 3.146\ldots$ [51, p. 126]. This expression shows that Plato probably arrived at the formula as the arithmetical mean of the half-perimeters of the inscribed square $2\sqrt{2}$ and the circumscribed hexagon $2\sqrt{3}$. As these initial values, with $3.464\ldots$ and $2.828\ldots$, are somewhat weak approximations for π, and yet the end value is quite accurate, one is attempted to interpret this as a typical case of the non-mathematician striking it lucky.

13.2 Polygons

After the initial phase largely consisting of empirical approximations, the Greek mathematician Archimedes of Syracuse (287–212 BC) inaugurated the first theoretical phase in the mathematics of π. Using a new method, he attained a milestone around 250 BC when, for the first time, he systematically approximated the number π and produced upper and lower limits on its value. In his book *The Measurement of the Circle* he stated three theorems about the circle, the third of which goes as follows:

3. *The ratio of the circumference of any circle to its diameter is less than* $3\frac{1}{7}$ *but greater than* $3\frac{10}{71}$.

When expressed as a formula, the theorem states

$$3 + \frac{1}{7} > \pi > 3 + \frac{10}{71} \tag{13.6}$$

and places π between the limits $3.14084\ldots$ and $3.14285\ldots$, which are accurate to two decimal places[1].

"This marks the origin of the value $\pi = 3\frac{1}{7}$, which spread victoriously from country to country and from book to book", writes Tropfke [114, p. 273]. "In Alexandria it quickly replaced the old value of $(16/9)^2$, as it was just as convenient but more accurate. From Alexandria news of Archimedes's value for π spread to India, and it has even been demonstrated to have been known in distant China in the fifth century AD."

Archimedes's "paper" [6] has only survived in parts; nearly all the passages in the Doric dialect which Archimedes spoke have been lost. Nevertheless it has been possible to largely reconstruct Archimedes's proof.

To prove his theorem, Archimedes calculated the perimeters of two regular 96-sided polygons such that one of them touched the circle from the outside and the other from the inside. The perimeter of the circumscribed polygon is slightly larger than the circumference of the circle and thus provides an upper limit for π, while the perimeter of the inscribed polygon is slightly smaller, yielding a lower limit for π.

Archimedes's 96-sided polygons were obtained by calculating a sequence of polygons in which each successive polygon had twice as many sides as its predecessor. He began with hexagons, whose geometric properties had been researched in-depth in ancient Greece. From hexagons, Archimedes then moved on to 12-sided polygons, and from there to 24-, 48-and finally 96-sided polygons in a total of four similar steps. The method of calculation entails first of all determining all the circumscribed polygons and then all the inscribed polygons, rather than first calculating the two hexagons and then the two 12-sided polygons etc. in criss-cross manner. Sometimes it is described otherwise.

Archimedes obtained the necessary formulae from the figures below:

---

[1] Today the left- and right-hand limits are generally transposed so that the theorem is written as $3 + \frac{10}{71} < \pi < 3 + \frac{1}{7}$, but in Archimedes's time it was customary to write it as shown above; moreover it was the upper limit which he calculated first.

Circumscribed polygons <space> Inscribed polygons

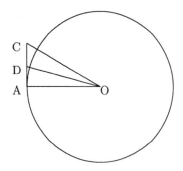 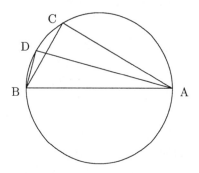

$(CO + OA) : CA = OA : AD$ <space> (13.7) <space> $(AB + AC) : BC = AD : DB$ <space> (13.9)
$OA^2 + AD^2 = DO^2$ <space> (13.8) <space> $AD^2 + DB^2 = AB^2$ <space> (13.10)

The labels in the pictures are the same as those used by Archimedes. Lines AC and BC denote the half sides of a circumscribed or inscribed polygon with n-sides, AD and BD denote the half sides of the "succeeding" $2n$-sided polygon; OD and AD thus have the effect of halving angles AOC and BAC.

The formulae (13.7) and (13.9) can be derived from elementary geometry, but this is not a trivial task. Archimedes had to calculate them without trigonometric half-angle formulae, and with little in the way of formulae to refer to.

The formulae, which enable the transition from polygons of n sides to polygons of $2n$ sides, gave Archimedes in the circumscribed as in the inscribed case two completely symmetrical sequences of numbers of $a_6, a_{12}, a_{24}, \ldots, a_{96}$ and $b_6, b_{12}, b_{24}, \ldots, b_{96}$, which are related to each other as follows:

$$a_{2n} = a_n + b_n \qquad (13.11)$$

$$b_{2n} = \sqrt{a_{2n}^2 + c^2} \qquad (13.12)$$

Writing this using our standard trigonometric notation, which Archimedes did not yet possess, this implies $a_n = c/\tan \alpha$, $b_n = c/\sin \alpha$, $\alpha = \pi/n$, where c is an arbitrary scaling constant which helps to make the calculations easier. Archimedes's two formulae (13.11) and (13.12) are probably the first recursion formulae in history and inevitably propelled him from the hexagon to the 96-sided polygon.

After deriving these formulae, Archimedes "only" needed to input an initial approximation and he could then start calculating. The starting ratios $OA : AC$ and $AC : CB$ come in both hexagons to $= \sqrt{3}$, and the accuracy of all subsequent calculations depends on the accuracy of this $\sqrt{3}$. As Archimedes had no other means at his disposal, he had to use two approximations, and without a word of explanation he chose these two:

$$\frac{1351}{780} > \sqrt{3} > \frac{265}{153} \qquad (13.13)$$

Scarcely any puzzle has so exercised the brains of mathematics historians as the question of how Archimedes could have arrived at these preferred values, which are only $2 \cdot 10^{-5}$ and even $5 \cdot 10^{-7}$ away from the true value, which is $1.7320508\ldots$. The simplest explanation [64, II, p. 51] is that he obtained them from the inequalities

$$a + \frac{b}{2a} > \sqrt{a^2 + b} > a + \frac{b}{2a+1} \qquad (13.14)$$

which were known long before his time, probably already in the time of the Babylonians. In the above formulation a^2 is the next lower or next higher perfect square below or above the radicand $a^2 + b$. This formula (13.14) produces the following on the first occasion that $a = 2$ and $b = -1$ are used:

$$2 - \frac{1}{4} > \sqrt{3} > 2 - \frac{1}{3}$$

$$\frac{7}{4} > \sqrt{3} > \frac{5}{3}$$

Of the two limits, the left one provides a better approximation but the right one contains the smaller numbers and therefore Archimedes could have chosen 5 as an approximation for $3\sqrt{3} = \sqrt{27}$ and used 13.14 once again.

$$5 + \frac{2}{10} > 3\sqrt{3} > 5 + \frac{2}{11}$$

$$\frac{26}{15} > \sqrt{3} > \frac{19}{11}$$

If he now used the 26 from the left-hand side as an approximation for $\sqrt{3 \cdot 15^2} = \sqrt{675}$, then after applying (13.14) again he could have arrived directly at 13.13 via

$$26 - \frac{1}{52} > 15\sqrt{3} > 26 - \frac{1}{51}$$

$$\frac{1351}{780} > \sqrt{3} > \frac{1325}{765} = \frac{265}{153}$$

In any case, Archimedes took the numerators 1351 and 265 from (13.13) for the a_6, the denominators 780 and 153 for the c, and $2 \cdot c$ (from "Pythagoras") for the b_6. He then performed four iterations for each of the 12-, 24-, 48-, and 96-sided polygons. In each iteration he had to approximate a square root using a fraction, and for this he evidently used the same effective approximation method (13.14) that he had used for the initial $\sqrt{3}$. In any case the basic inaccuracy of the area of the 96-sided polygon instead of the area of the circle remained greater by factors (8 and 3) than that of the calculation. The mathematical procedure, most of which has survived (only the expressions in square brackets have been added) can be represented as follows [64, II, p. 55]:

Circumscribed polygons				Inscribed polygons		
a	b	c	n	a	b	c
265	306	153	6	1351	1560	780
571	$> [\sqrt{571^2 + 153^2}]$ $> 591\frac{1}{8}$	153	12	2911	$< \sqrt{2911^2 + 780^2}$ $< 3013\frac{1}{2}\frac{1}{4}$	780
$1162\frac{1}{8}$	$[\sqrt{(1162\frac{1}{8})^2 + 153^2}]$ $> 1172\frac{1}{8}$	153	24	$5924\frac{1}{2}\frac{1}{4}$ 1823	$\ldots$ $< \sqrt{1823^2 + 240^2}$ $< 1838\frac{9}{11}$	780^1 240
$2334\frac{1}{4}$	$[\sqrt{(1162\frac{1}{8})^2 + 153^2}]$ $> 2339\frac{1}{4}$	153	48	$3661\frac{9}{11}$ 1007	$\ldots$ $< \sqrt{1007^2 + 66^2}$ $< 1009\frac{1}{6}$	240^1 66
$4673\frac{1}{2}$		153	96	$2016\frac{1}{6}$	$< \sqrt{(2016\frac{1}{6})^2 + 66^2}$ $< 2017\frac{1}{4}$	66

[1]Here the fractions of a and c are reduced to simpler expressions.

At the end Archimedes had, with $c_{96} : a_{96}$ and $c_{96} : b_{96}$, calculated approximations for the ratio of polygon side to diameter whose relative error was less than $4 \cdot 10^{-5}$ both for the circumscribed and the inscribed 96-sided polygons. Through multiplication of the fractions $c_{96} : b_{96}$ and $c_{96} : a_{96}$ by 96 he arrived at

$$\pi < \frac{96 \cdot 153}{4673\frac{1}{2}} = 3 + \frac{667\frac{1}{2}}{4673\frac{1}{2}} < 3 + \frac{667\frac{1}{2}}{4672\frac{1}{2}} = 3 + \frac{1}{7} \qquad (13.15)$$

and

$$\pi > \frac{96 \cdot 66}{2017\frac{1}{4}} = \frac{6336}{2017\frac{1}{4}} = 3 + \frac{284\frac{1}{4}}{2017\frac{1}{4}} > 3 + \frac{10}{71} \qquad (13.16)$$

13.2 Polygons

and thus completed the famous proof of his famous theorem.

Voilà! What is the most remarkable thing about Archimedes's π calculation? The accuracy of the result? The fact that it produced *two limits* for π, which was the first indication that there is no simple value for π? Or the method of calculation itself, which was quite novel in its manner and scope?

Archimedes's book, *The Measurement of the Circle* – actually a misnomer as it is in fact concerned with *calculations* rather than measurements – contains two additional theorems.

The first of these states that every circle has the same area as a right-angled triangle one of whose sides is equal to the radius and whose hypotenuse is equal to the circumference. This theorem Archimedes proved indirectly by showing that the area of the circle cannot be either greater or smaller than the area of such a triangle, hence it must be equal to the triangle. His second theorem states that the ratio of the circle to the square of its diameter is more or less 11 : 14. This is obvious from the third theorem stated above, as is shown by the relationships $3 + \frac{1}{7} = \frac{22}{7} = 4 \cdot \frac{11}{14}$.

Archimedes's approximation method can in principle be used to calculate π to *any* degree of accuracy, and in fact up to the 17th century many of his successors did nothing more than calculate polygons with ever greater numbers of sides. In any case, his method remained without a serious rival for nearly 2000 years.

Unfortunately, however, Archimedes's polygon method does not converge particularly well. Even when the lengths of the sides of circumscribed and inscribed 96-side polygons are calculated exactly, the absolute error of the π approximation thus achieved always remains $0.00112\ldots$ or $-0.00056\ldots$. Every time the number of sides is doubled, the error reduces only by around 1/4. When formulated accurately, regular polygons with n sides produce (at least) $\lfloor 2 \log_{10} n - 1.19 \rfloor$ correct decimal digits. In this way, for example, 10^{18}-sided polygons are required to calculate π to just 36 places. This approach thus guaranteed a lot of sweat.

Archimedes himself seems to have improved his calculation. Heron (1st century AD) refers to a work by Archimedes which has since been lost, in which the limits are stated at $211875 : 67441 = 3.14163\ldots$ and $197888 : 62351 = 3.1738\ldots$ [114, p. 273]. But however accurate the first value is, there must be something wrong overall as both values are greater than π.

400 years after Archimedes's time the Greek astronomer Ptolemy (150 AD) also came up with the approximation 3.1416. It may be that he discovered this approximation by simply deriving the approximate value $3°8'30'' = 3\frac{17}{120}$ from the sexagesimal representations of Archimedes's limits, $3\frac{1}{7} = 3°8'34.28''$ and $3\frac{10}{71} = 3°8'27.04''$ [114, p. 275], or else he may have separately summed the numerators and denominators from their fractions $3 + 1/7 = 154/49$ and $3 + 10/71 = 223/71$ to arrive at the mean value of $377/120$ [51, p. 127].

Rome

As far as we know, the Romans did not make any further progress in the mathematical analysis of circles. As late as 15 BC no more precise a value than $\pi = 3$ appears to have been either required or known. Surveying in Roman times was evidently not particularly accurate [114, p. 277].

China

When Wang Mang came to the throne at the end of the Western Han dynasty (206 BC–24 AD) he commissioned the astronomer and calendar expert Liu Xin to develop a standard measurement for his empire. Liu Xin's response was to produce a cylindrical vessel made of bronze. It is assumed that around 100 copies of this were made and distributed throughout the country. One of these has been preserved in the Peking Palace Museum. From analysing it, historians have concluded that Liu Xin must have possessed a π approximation which was significantly better than 3 and may even have been the accurate value of 3.1547. This would then have been around the beginning of the Common Era.

This and also everything else we know about the history of π in China comes from an paper [78] by Lam Lay-Yong and Ang Tian-Se, most of which has been confirmed by [51].

Around a century later Zhang Heng (78–139 AD) improved the value of π experimentally in the following manner. In his day the ratio of the area of a square to an inscribed circle was assumed to be $4:3$. By analogy, Zhang Heng calculated that the ratio of the volume of a cube to that of an inscribed sphere was its square, i.e. $4^2 : 3^2 = 16 : 9$.

As he probably realised that this was too large, he corrected it to 8 : 5. When calculating in the other direction, to be consistent he took the root $\sqrt{8} : \sqrt{5}$ for the circle : square ratio and from this he calculated π at $\sqrt{10} = 3.1622\ldots$.

In the third century, Wang Fan (217–257 AD) came up with the circumference : diameter ratio of $\frac{142}{45} = 3.155\ldots$.

He was followed by two important scholars whose π calculations were systematic and produced better results. The first of these was Liu Hui c. 263 AD. He started with a circle of radius 10 and, using Pythagoras's theorem, he calculated the areas of inscribed polygons starting with the hexagon and proceeding upwards to polygons with 192 sides. His calculation ended with the following inequality:

$$314\frac{64}{625} = A_{192} < A < A_{96} + 2(A_{192} - A_{96}) = 314\frac{169}{625} \qquad (13.17)$$

Here, A refers to the area of the circle and A_n to the area of an inscribed polygon with n sides.

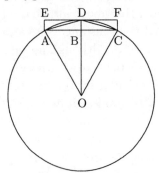

Liu Hui obtained his upper limit from the inscribed polygon with double the number of sides, unlike Archimedes who had obtained his upper bound from the circumscribed polygon with the same number of sides. The expression $2(A_{2n} - A_n)$ means n times rectangle AEFC.

At the end of his calculation Liu Hui then established the following approximation from his two limits, which differ by $\frac{105}{625}$:

$$A \approx A_{192} + \frac{36}{625} \qquad (13.18)$$

From this he then arrived at the following approximation for π

$$\pi \approx \frac{314 + \frac{4}{25}}{10^2} = 3.1416 \qquad (13.19)$$

Liu Hui, 263 AD

which differs from the true value of π by $7.346 \cdot 10^{-6}$.

Liu Hui did not give any justification for the relationship (13.18), so we can only guess as to how he arrived at it. No doubt it was clear to him that the true value he was looking for was closer to the 192-sided polygon than to the 96-sided polygon, so he probably used a weighted mean instead of the arithmetic mean. In this way he may have divided the difference $169 - 64 = 105$ by 3 instead of halving it, which would have resulted in the approximation $314 + 99/625$ (which possesses almost the same error, but with the opposite sign). We can well understand how the scholar was unwilling to leave things there: the fraction $\frac{99}{625}$ "is crying out" for a small manipulation which would enable it to be replaced by the fraction $\frac{100}{625}$, which is much easier to shorten and produces $\frac{4}{25}$. This was the final crowning touch to a formula (13.19) which is both accurate and easy to use.

Liu Hui's method is similar in many respects to that of Archimedes (see pages 171ff.): both mathematicians used the exhaustion method applied to polygons and started with a hexagon, proceeding from there in a number of stages in which the number of sides was doubled. And both men were the only π mathematicians of ancient times to calculate *two values* for π, namely an upper and a lower limit.

Unlike Archimedes, Liu Hui only used inscribed polygons and calculated their areas rather than their perimeters. But above all, he used a decimal place system which he had inherited from his ancestors and which by his day already possessed the decimal zero. Although he subsequently reverted to fractions, he was able to dispense with fractions during the calculations. This meant that the calculation process was significantly simpler and serves as a particularly good example of the capability of the Chinese in ancient times to handle large numbers. Liu Hui was, for example, aware that he needed to calculate radicands from square roots to 10 decimal places accurately in order to obtain 5 correct decimal places in the result. Nevertheless, the high accuracy of (13.19) is not due to his use of decimal numbers but to the fact that the mean of the upper and lower limit was almost optimally weighted. (That this was the case was first established in 1621 by Wildebrod Snell, see page 183.)

The second important ancient Chinese π scholar was Tsu Chhung-Chih (or Zu Chongzhi) (429–500). He improved the precision of π from Liu's value by around 2 powers of ten to the limits

$$3.1415926 < \pi < 3.1415927 \qquad (13.20)$$

This interval, with its 7 accurate decimal places, held the world record for 800 years, although to all intents it remained a secret as even in China the two limits were not rediscovered until the fourteenth century.

Another interesting feature of the result is that Tsu Chhung-Chih was one of the first to express these numbers using decimal notation, when he wrote [116, p. 312]:

3 zhang, 1 chi, 4 cun, 1 fen, 5 li, 9 hao, 2 miao, 7 hu

In this expression, zhang, chi etc. are all units of length which behave like $1 : 10 : 100 : \ldots$.

If Tsu used the same method of calculation as Liu, as seems likely, he used polygons with 12,288 sides to calculate π and had to record his interim results with up to 13 decimal places.

Tsu also found a second value which was numerically less accurate but more visually attractive. This is as follows:

$$\pi \approx \frac{355}{113} = 3.1415929\ldots \qquad (13.21)$$

This decimal fraction, which is accurate to 6 decimal places, agrees with the fourth partial quotient of the simple continued fraction of π (see page 54), but it was definitely not obtained in this way. It is assumed that he obtained the fraction from two less accurate values which were then combined fortuitously to produce an especially good mean.

India

In 499 AD the astronomer Aryabhata (born 476 AD) wrote a work called *Aryabhatiya*. His theorem 10 (out of 133 theorems) goes as follows [72, p. 203]:

> "Add 4 to 100, multiply by 8 and add 62,000. This is the approximate circumference of a circle of which the diameter is 20,000."

This produces a value for π of $62832/20000 = 3.1416$. However, this value is suspected of being of Greek origin, possibly even originating from Archimedes, and of having found its way to India like many other things. But Aryabhata may have calculated it himself, for he demonstrated that if a is the length of the side of a polygon with n sides which is inscribed in a circle of diameter 1, and b is the length of a side in a polygon with twice as many sides $2n$, then $b^2 = (1 - \sqrt{1-a^2})/2$. Beginning with a hexagon he calculated the side lengths of polygons with $12, 24, \ldots, 384$ sides. The perimeter of the last polygon he stated as $\sqrt{9.8694}$, and from this he calculated the previously mentioned value of 3.1416 by approximation of this root [14, p. 341].

The Hindu Brahmagupta (born 598 AD) was fascinated by the discovery that the perimeters of regular polygons with 12, 24, 48 and 96 sides and diameter 10 are $\sqrt{965}$, $\sqrt{981}$, $\sqrt{986}$ and $\sqrt{987}$. He concluded from this that if the number of sides were doubled again, the perimeter would tend towards $\sqrt{1000}$. Using this logic he arrived (c. 650 AD) at $\pi = \sqrt{1000}/10 = \sqrt{10} = 3.16227\ldots$ [41, p. 68].

Choresmia (today Uzbekistan)

In c. 830 the Choresmian Alkarism made his mark on the history of π by coming up with 3 values of π, 22/7, $\sqrt{10}$ and 62832/20000. The first of these was intended to serve as a mean value, the second was meant for geometricians and the third for astronomers. We also have Alkarism or Abu Ja'far Mohammed ibn Mûsâ al-Khowârizmî to thank for our present-day word algorithm, which is based on his name. He wrote a famous book entitled *Kitab al jabr w'al-muqabala*; and it is from the title of that work that our word algebra [77, p. 197] is derived.

Middle Ages

In Europe in around 1220 Leonardo of Pisa (1180–1240?), better known by the name *Fibonacci*, independently calculated the value $864/275 = 3.14181\ldots$ from a 96-sided polygon, without drawing on Archimedes's work. His approach was more accurate than Archimedes's, but not as systematic, so according to Tropfke his 3 correct decimal places owe more to chance than to his calculation method.

It is said that Dante Alighieri (1265–1321), author of "The Divine Comedy", possessed the approximation $\pi = 3 + \sqrt{2}/10 = 3.14142\ldots$ [41, p. 68].

At any rate it was Dante who composed that secular verse which could aptly summarise all the endeavours of π researchers[2]:

> As the geometer his mind applies
> To square the circle, nor for all his wit
> Finds the right formula, howe'er he tries,...

---

[2] Paradise, Canto XXXIII, lines 133-135, translated by Dorothy Sayers and Barbara Reynolds.

13.2 Polygons

Nicely put indeed. But unfortunately in this stanza he is not talking about the calculation of π but about the "squaring the circle", which in Dante's day was still viewed as an unresolved problem rather than one that is intrinsically insoluble.

Leonardo da Vinci (1452–1519) came up with a "mechanical" approximation. The inventor of many machines once rolled a circular cylinder whose height was equal to half the radius of the cross-section for one revolution. The area covered while such a cylinder is rolling takes the form of a rectangle and can be measured easily. It is equal to the area of the cross-section of the cylinder, i.e. the area of a circle, πr^2 [26, p. 27].

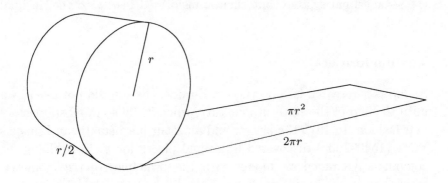

All in all, the European Middle Ages saw little progress in the matter of π. Thus in 1464 the man regarded as the greatest European mathematician of the fifteenth century, Regiomontanus (1436–1476), whose real name was Johannes Müller, used a value of 3.14243 for π. And Adrianus Metius (1571–1635), rated equally important, was using nothing better than Tsu Chhung-Chih's approximant 355/113 which the Chinese mathematician had discovered 1000 years earlier, and even this he only found through a happy accident when he calculated the mean values of the numerators and denominators of two limits (377/120 and 333/106) [14, p. 343]. At a time when the mathematicians themselves could not produce anything better than this, we should praise the approximation of $\pi = 3\frac{1}{8} = 3.125$ which the artist Albrecht Duerer (1471–1528) used, 3500 years after the Babylonians had first invented it. This value can at least be well constructed with a compass and ruler.

By far the best approximation to π for 180 years was calculated by the Persian astronomer of the observatory in Samarkand, Al-Khashî, c. 1430. He calculated the following value for 2π using polygons with $3 \cdot 2^{28}$ sides [77, p. 198]:

$$2\pi = 6.28318\,53071\,79586\,5$$

This value is accurate to 16 decimal places.

While Al-Khashî may in principle have used the same method as Archimedes, he performed his calculations using a different approach [69, pp. 314–319]. Especially impressive is the way he continuously monitored his results for errors, which allowed him to ignore redundant decimal places. Al-Khashî performed his calculations using the sexagesimal system, but then transformed his 2π to the decimal system.

The modern era

As far as π research is concerned in Europe, the sun did not rise again until the end of the sixteenth century, when it did so in most spectacular fashion. In 1579 the lawyer and amateur mathematician François Viète (1540–1603) discovered 9 decimal places for π. In so doing, he combined Archimedes's model with the principles of trigonometry, which had in the meantime become established, to calculate polygons with $3 \cdot 2^{17} = 393\,216$ sides. His results are as follows [114, p. 284]:

$$n \cdot \sin \frac{180}{n} < \pi < n \cdot \tan \frac{180}{n}$$

and with $n = 393216$,

$$3.1415926535 < \pi < 3.1415926537$$

To find the values of the sine and tangent, Viète applied the formula $2\sin^2 \theta/2 = 1 - \cos\theta$ [14, p. 343] iteratively.

Several other Dutchmen then went on to leave their mark on the history of π. The first of these was Adrian van Romanus (1561–1615), who in 1593 achieved distinction by calculating polygons with one billion sides in order to arrive at 15 accurate decimal places. Three years later, in 1596, he was surpassed by Ludolph van Ceulen (1539–1610), a professor of mathematics at the University of Leyden. At first Ceulen calculated only 20 digits, for which he used the perimeters of the inscribed and circumscribed regular polygons of $60 \cdot 2^{33} \approx 480$ *billion* sides.

He continued his calculations afterwards, but did not publish them himself. A posthumous publication from the year 1615 specifies 32 decimal places derived from polygons of $2^{62} \approx 4 \cdot 10^{18}$ sides and in 1621 as many as 35 places were attributed to him. These 35 places were very likely carved on his tombstone (which has since been lost). But Ceulen's record lived on, with the result that up to the time of the First World War π was still known in Germany as the "Ludolphian number".

Shortly before the polygon method of calculating π was replaced by the series method it underwent yet one last significant improvement. In particular, Wildebrod Snell (1581–1626) succeeded in 1621 in obtaining 34 digits of π from 2^{30}-sided polygons, whereas Ludolph van Ceulen, using the same method as Archimedes, had only managed 16. Snell divided every sector of the polygon one more time by 3 and constructed from each polygon side two further sides which contained the arc more accurately. His method was based on the inequality [14, p. 344]

$$\frac{3\sin\theta}{2+\cos\theta} < \theta < 2\sin\frac{\theta}{3} + \tan\frac{\theta}{3} \tag{13.22}$$

The lower limit had already been discovered by the German cardinal Nicolaus Cusanus (1401–1464) [114, p. 281], while the upper limit was discovered by Snell himself. The accuracy of the procedure can be easily inferred today from the series expansions of sine and tangent. For it appears that the third order terms in the expression $2\sin x + \tan x$ drop out, resulting in a quadratic error improvement. Moreover, the upper limit of Snell's inequality (13.22) confirms the correctness of Liu Hui's (intuitive) approach in his calculation of π 1400 years earlier (see page 177).

It was using Snell's method that Grienberger, the last mathematician of the polygon era, which had lasted almost 2000 years, calculated π to 39 decimal places in 1630.

The seventeenth century saw the discovery of other alternatives to Archimedes's method. Whereas under Archimedes's method, first of all the perimeters of all the circumscribed polygons are calculated and only then are the perimeters of all the inscribed polygons calculated, the new approaches involved working in "criss-cross" fashion: first of all a pair comprising one inner and one outer polygon is calculated, each possessing n sides, and then a similar pair of polygons with $2n$ sides is calculated, followed by a further pair with $4n$ sides etc.

Such an approach was developed by James Gregory (1638?–1675), whom we will meet again as one of the originators of the Leibniz series (see page 188). Continuing in the style of Archimedes, he held the radius of the circle constant over all the iterations, but calculated – in criss-cross fashion – the areas I_n and i_n of the circumscribed and inscribed polygons instead of the perimeters. Beginning with a square and the radius $\sqrt{2}$, he repeated his procedure, which results in linear convergence on π, as shown below:

$$i_4 = 2, I_4 = 4; \quad i_{2n} = \sqrt{i_n \cdot I_n}, \quad I_{2n} = \frac{2 \cdot i_{2n} \cdot I_n}{i_{2n} + I_n} \qquad (13.23)$$

Gregory, c. 1667

The left-hand formula had already appeared in the Middle Ages, in the writings of Jordanus Nemorarius (–1237?), but Gregory is unlikely to have known of Nemorarius's work.

Another innovation was subsequently made by René Descartes (1596–1650), who came up with an idea which Cusanus had already tried 200 years earlier, but without success: every time one moves on to the next iteration, it is no longer the radius of the circle that is held constant but the perimeter of the polygons. From a square of perimeter P an octagon with an equal perimeter is constructed, and from this in turn a 16-sided polygon with an equal perimeter etc. [114, p. 289]. As the number of sides increases, the radiuses of the circumscribed and inscribed circles in each case converge from above and below towards the common value of $P/(2\pi)$. Beginning with the square and the perimeter $P = 8$, Descartes's recursion rule runs as follows:

$$R_4 = \sqrt{2}, r_4 = 1; \quad r_{2n} = \frac{1}{2}(r_n + R_n), \quad R_{2n} = \sqrt{r_{2n} \cdot R_n} \qquad (13.24)$$

Descartes, c. 1649

Descartes's procedure converges linearly towards the value $4/\pi$.

At this point we shall take a detour 160 years forward in time to 1809, when Carl Friedrich Gauss (1777–1855) came up with the following, almost identical recursion rule which also converges towards π, but a lot more quickly[3]:

$$R_4 = \sqrt{2}, r_4 = 1; \quad r_{2n} = \frac{1}{2}(r_n + R_n), \quad R_{2n} = \sqrt{r_n \cdot R_n} \qquad (13.25)$$

Gauss, 1809

---

[3] Strictly speaking, the R_{2n} and r_{2n} converge towards the value $\sqrt{f_{2n} \cdot \pi}$ with $f_4 = 1/2$; $f_{2n} = f_n - \frac{n}{4}(R_{2n}^2 - r_{2n}^2)$. This results from the formula (7.1) on page 87

It is difficult to believe, but the small difference of r_n instead of r_{2n} in the formula on the right-hand side actually results in quadratic convergence. The recursion no longer produces a constant improvement on every step but instead it delivers each time twice the improvement achieved during the previous step, yielding, for example, not just 6 accurate decimal places of π after 10 recursion steps but over 1000. A mathematical miracle!

Whereas Descartes's method is based on elementary geometry, Gauss's iteration rule is derived from the quite different theory of elliptic functions. The fact that the notation r and R is used in both formulae should not be taken to mean that the Gauss formula refers, like the Descartes formula, to the radiuses of circles.

The details, background and the story of how the Gauss algorithm came about are discussed in depth in Chapter 7, where the regrettable historical accident which resulted in this algorithm being overlooked for one-and-a-half centuries, until 1976, is also discussed.

13.3 Infinite expressions

The second era in the calculation of π began with infinite products and infinite series (sums). After almost 2000 years during which mathematicians had had only geometric methods at their disposal, these provided them at last with an analytical tool.

Beginnings in India

The first series for π were noted down in India in the fifteenth century, possibly even earlier. They are thus at least 100 years older than the first European series (1593). This fact might have been known for 100 years, but the essay by Charles M. Whish on this subject dating from the year 1835 *On the Hindu quadrature of the circle and the infinite series of the proportion of the circumference to the diameter exhibited in the four Sastras, Tantra Sangraham, Yukti-Bhasa, Karana-Paddhati and Sadratnamala* was overlooked for 100 years. It was not until around 1940 that this essay resurfaced and its essentials were confirmed.

No less than eight π series are found in the Sanskrit texts *Yukti-Bhasa* and *Yukti-Dipika*, among them, for example, the two following series, the first of which begins like the Leibniz series:

$$\frac{\pi}{4} \approx 1 - \frac{1}{3} + \frac{1}{5} - \frac{1}{7} + \cdots \mp \frac{1}{p-1} \pm \frac{p/2}{p^2+1} \qquad (13.26)$$

Nilakantha, 15th century [89]

$$\pi = 3 + \frac{4}{3^3-3} - \frac{4}{5^3-5} + \frac{4}{7^3-7} - \frac{4}{9^3-9} + \cdots \qquad (13.27)$$

Nilakantha, 15th century [89]

All eight of the series are included in our collection of formulae, items (16.6) to (16.13).

We are indebted to S. Parameswaran [89] and Rinjan Roy [98] for making us aware of them. According to Parameswaran, the origins of these series are complicated, especially as they were passed down orally. The Yukti-Dipika was composed by Sankaran (c. 1500–1560), who claims to have received directions from Jyestha-devan, who for his part wrote down the *Yukti-Bhasa*. This contains detailed proofs for the series as well as the series themselves. These two scholars in turn received instruction in astronomy and mathematics from Kelallur Nilakantha Somayaji (c. 1444 – c. 1545) [sic!], the author of the work *Tantra Sangraham*. If the series were not already composed, then the latest person who could have written them is Nilakantha. Recent research suggests that some of the series could actually have been formulated earlier and, in particular, some of them are linked with the name of Madhavan, who lived from c. 1340 to 1425.

Even more surprising, if that is possible, than the great age of these Indian series is the fact that they are quite simply better than the first European series.

The author Nilakantha provided a residual sum along with the series (13.26). This gives an estimate of the order of magnitude of p^{-5}. (See formula (12.5)). As far as we know, Europe had to wait another hundred years for an estimate that would be as accurate as this, until Euler produced one in 1750.

The second series (13.27) is significantly better than the first (13.26) as regards convergence. For example, the first series required 10^{10} terms to produce 10 correct decimal places of π, whereas the second required fewer than 10^3.

A third series (16.6) is even more convergent, in that 28 terms produce as many as 16 accurate decimal places.

Infinite products

In Europe, François Viète, whom we mentioned above, produced an infinite expression, or, to be more precise, an infinite product. in 1593 he achieved this feat, remarkable both historically and aesthetically, in his work *Variorum de Rebus Mathematicis* [115]:

$$\frac{2}{\pi} = \sqrt{\frac{1}{2}} \cdot \sqrt{\frac{1}{2} + \frac{1}{2}\sqrt{\frac{1}{2}}} \cdot \sqrt{\frac{1}{2} + \frac{1}{2}\sqrt{\frac{1}{2} + \frac{1}{2}\sqrt{\frac{1}{2}}}} \cdots \quad (13.28)$$

Viète, 1593

The product is also numerically interesting as it converges very quickly to π, generating 15 accurate decimal places after only 25 terms.

Viète arrived at his formula through a simple but cunning trick. He applied the standard formula $\sin(x) = 2\sin(x/2)\cos(x/2)$ recursively and thus obtained $\sin(x) = 2^n \sin(x/2^n) \cos(x/2) \cdots \cos(x/2^n)$. As n becomes bigger, $2^n \sin(x/2^n)$ converges to x. To calculate the cos expressions, Viète used another standard formula $\cos(x/2) = \sqrt{\frac{1}{2}(1 + \cos x)}$. He then only needed to substitute $\pi/2$ for x, i.e. to set $\sin(x) = 1$ and $\cos(x) = 0$, and - hey presto! - there was his clever formula.

Curiously in the history of π, the Viète product was followed by another product. This is curious because apart from this product and Viète's product, there are few other infinite products for π but a large number of infinite *series*. In any case in 1655 John Wallis (1616–1703) reported the following infinite product in his *Arithmetica Infinitorum*, derived by successive interpolations:

$$\frac{4}{\pi} = \frac{3 \times 3 \times 5 \times 5 \times 7 \times 7 \times 9 \times 9 \times 11 \times 11 \times 13 \times 13 \cdots}{2 \times 4 \times 4 \times 6 \times 6 \times 8 \times 8 \times 10 \times 10 \times 12 \times 12 \times 14 \cdots} \quad (13.29)$$

Wallis, 1655

Of course no mathematics student today would be allowed to write the product in this way because it results in $\frac{\infty}{\infty}$. To show the unconditional convergence, two numerators and two denominators must each be combined together: $\frac{4}{\pi} = \frac{3 \cdot 3}{2 \cdot 4} \cdot \frac{5 \cdot 5}{4 \cdot 6} \cdots$ [114, p. 292].

One can say that Wallis computed the Integral $\int_0^1 x^m y^k dx$ (with $x^2 + y^2 = 1$) for integral values of m and k [20, p. 68–80]. Wallis's derivation is historically significant as it inspired Newton to a similar approach which ultimately led him to the binomial series.

Even if the Wallis product can be calculated easily, it cannot be used for effective π calculations because after 100 terms, for example, it is only accurate to $7.8 \cdot 10^{-3}$.

William Viscount Brouncker (c. 1620–1684) was the first President of the Royal Society and "manipulated" (Beckmann) the Wallis product into a continued fraction for π, the first in the history of π, in 1658:

$$\frac{4}{\pi} = 1 + \cfrac{1^2}{2 + \cfrac{3^2}{2 + \cfrac{5^2}{2 + \cdots}}} \tag{13.30}$$

Brouncker, 1658

According to Beckmann [18, p. 131] we can only guess as to how Brouncker arrived at this continued fraction. Leonhard Euler was the first person to derive it systematically in 1775, i.e. over 100 years later, from the series expansion of arctan 1.

At the time when the first series were discovered, infinitesimal analysis did not yet exist. This explains why the mathematicians concerned found the derivations so difficult.

Infinite sums

The independent discoveries by Isaac Newton (1643–1727) and Gottfried Wilhelm Leibniz (1646–1716) of differential and integral calculus in the second half of the 17th century heralded a boom in series formulae for π.

Some time in 1665–1666, Isaac Newton himself succumbed to the temptation to calculate π with the aid of one of the new series. He used his *Method of Fluxions and Infinite Series* to derive a quite effective series for π (see (16.63)) which is essentially an expansion based on the arcsin function. Using this, he obtained 15 digits of π (which, of course, was no world record at that time). Newton later wrote, "I am ashamed to tell you to how many figures I carried these computations, having no other business at the time" [32, p. 339]. His words show only too well the two minds which characterise many π "digit hunters".

The date 15 February 1671 is of historical importance as the date on which James Gregory (1638?–1675) stated the first series for arctan (without explaining how he derived it), in a letter [20, p. 89]. If r is the radius, a the arc and t the associated tangens, then the *Gregory series* reads as follows [114, p. 293]:

$$a = t - \frac{t^3}{3r^2} + \frac{t^5}{5r^4} - \frac{t^7}{7r^6} + \frac{t^9}{9r^8} - \ldots \qquad (13.31)$$

which we write in modern form ($r = 1$):

$$\arctan x = x - \frac{x^3}{3} + \frac{x^5}{5} - \frac{x^7}{7} + \frac{x^9}{9} - \ldots \qquad (13.32)$$

Gregory himself did not perform the easy step of substituting 1 for x which leads to the so-called *LeibnizSeries*:

$$\frac{\pi}{4} = 1 - \frac{1}{3} + \frac{1}{5} - \frac{1}{7} + \frac{1}{9} - \ldots \qquad (13.33)$$

Leibniz derived this series independently of Gregory and cited it in letters to several friends from 1674 onwards. In the meantime we know that the same series was already known much earlier in India (see page 185).

The first person to make use of the Gregory series to obtain an approximation for π was the astronomer Abraham Sharp (1651–1742). In (13.32) he substituted $\tan(\pi/6) = \sqrt{1/3}$ for x and in this way discovered a series (see (16.67)) which produces about 1 digit per 2 terms. This he used in 1699 (almost 40 years after Gregory's invention) to calculate π relatively easily to 71 correct decimal places. This was the very first world record that was not based on Archimedes's polygon method.

Leonhard Euler

18th century mathematics was dominated by the Swiss mathematician Leonhard Euler (1707–1783). His almost inconceivably extensive life's work, consisting of over 700 books and essays, which take up 80 (still not completely published) volumes, attests to his status as the most versatile and prolific mathematician in the history of mathematics. His creativity and productivity were not impaired in the least by the persistent problems he suffered with his eyes, which resulted eventually in complete loss of his sight.

Even though he is sometimes identified more with the number e, Euler has the distinction of having decisively increased our understanding of π. He came up with an exceptional number of formulae for

π, including the one which is commonly rated as the most fascinating of all time, because it unites in one formulae 5 basic quantities of mathematics:

$$e^{i\pi} + 1 = 0 \qquad (13.34)$$

You would not believe how much has already been said and written about this formula! In the introduction we quoted the enthusiastic words of Benjamin Peirce, but he was only one of many who have marvelled effusively at the formula. Only a few years ago (1990) it was voted the "most beautiful of all" by the readers of *Mathematical Intelligencer* [120][4]

Euler did not actually express "his" formula in this way but like this (1743, in *Miscellanea Berolinesia* [41, p. 89]):

$$e^{iz} = \cos z + i \sin z \qquad (13.35)$$

Euler, 1743

which through $z = \pi$ turns into (13.34) because $\cos \pi = -1$ and $\sin \pi = 0$. This expression was not entirely new. Roger Cotes (1682–1716) had already published the equivalent equation in 1714

$$\sqrt{-1}x = \log(\cos x + \sqrt{-1} \sin x) \qquad (13.36)$$

Cotes, 1714

albeit using a different notation [39, p. 481]. Finally, mathematicians like writing Euler's theorem somewhat cryptically, for example

$$\pi = -i \ln(-1) \qquad (13.37)$$

David Wells, commenting on the results of the reader survey quoted above, wondered whether small and "inessential" changes to the formula (13.34) could also change its aesthetic value and thus the degree of favour it would find among readers. For example, he wondered how the following equivalent formula would have scored:

$$i^{-i} = e^{\pi/2} \qquad (13.38)$$

Among Euler's most spectacular results in the π domain is his solution for the sum of the reciprocal perfect square roots, i.e. for the expression $\frac{1}{1} + \frac{1}{4} + \frac{1}{9} + \cdots + \frac{1}{n^2} + \cdots$. Many great mathematicians,

---

[4] In second place was another of Euler's formulae, namely that which defines the relationship between the number of vertices (V), the number of edges (E) and the number of faces (F) in a polyhedron: $V - E + F = 2$. in 14th place was an Indian π series dating from the 15th century (16.10).

among them Leibniz and the brothers Jakob (1654–1705) and Johann Bernoulli (1667–1748), had tried their hand at this apparently simple exercise without success. Euler was the first one to succeed, in 1736, in resolving this sum with the elegant expression,

$$\sum_{k=1}^{\infty} \frac{1}{k^2} = \frac{\pi^2}{6} = 1.64493\,406\ldots \tag{13.39}$$

By way of proof, Euler considered the equation

$$\sin x = 0 \tag{13.40}$$

For $\sin x$, he took the series that was already familiar to Newton

$$\sin x = x - x^3/3! + x^5/5! - x^7/7! + \ldots \tag{13.41}$$
$$= x\left(1 - x^2/3! + x^4/5! - x^6/7! + \ldots\right) \tag{13.42}$$

or, after substitution of $y = x^2$,

$$= x\left(1 - y/3! + y^2/5! - y^3/7! + \ldots\right) \tag{13.43}$$

In order for the right-hand side of the equation to $=0$, either x or the expression in brackets must $=0$. The known solutions of (13.40) $x = \pm\pi, \pm 2\pi, \pm 3\pi$ thus hold true for the expression in brackets where $y = \pi^2, 2^2\pi^2, 3^2\pi^2, \ldots$ correspond to them. Euler then used the theorem from algebra, which states that the sum of the inverse values of the n solutions of the polynomial $1 + a_1 x + a_2 x^2 + \ldots + a_n x^n$ is equal to $-a_1$, i.e. to the negative of the first coefficient. (Example: the equation $-\frac{1}{6}(x-1)(x-2)(x-3) = 0$ has the solutions 1, 2 and 3. The sum of their inverse values is equal to $\frac{11}{6}$ and this is also the negative coefficient of x, as one can establish by multiplying out the left-hand side of the equation.) Euler assumed that this must also apply to the *infinite* sum contained in the brackets, which has the same form. This first coefficient in (13.43), set negative, is $1/3! = \frac{1}{6}$; this value must thus equal the sum of the inverse values of the solutions

$$\frac{1}{1^2\pi^2} + \frac{1}{2^2\pi^2} + \frac{1}{3^2\pi^2} + \ldots = \frac{1}{6} \tag{13.44}$$

If both sides are multiplied by π^2, one obtains the sought-after solution to this unusual problem.

In Euler's previously mentioned primary work *Introductio in analysin infinitorum* of 1748, he delivered another proof for the sum (13.39), which satisfies more stringent requirements.

This work also contains "truckloads" (Beckmann) of other π formulae, e.g. similar summation expressions for π^3, π^4, π^6 etc. as well as many more beautiful identities. It would actually require a separate monograph to do justice to the wealth of Euler's discoveries on π. Nevertheless, a small selection of his pronouncements is included in our collection of formulae on page 224ff.

Another of Euler's achievements was, as mentioned above, to make the symbol π popular. Then, at the age of 72 in 1779, he began calculating π and arrived at 20 decimal places using an arctan formula he had developed himself and a series of his own for the arctan function (see page 74). This only took him an hour.

But Euler was not the only eighteenth century mathematician to add to our collective knowledge of π. Already in the 15th century it was suspected that π could not be a rational number, i.e. that it could not be represented as a quotient of two integers. Huygens and Leibniz were quite convinced of this. But it was not until 1766 that the Alsatian mathematician Johann Heinrich Lambert (1728–1777) produced the proof that π is irrational in [79], His proof begins with the statement of an infinite continued fraction for $\tan v$, which he expressed like this:

$$\tan v = \cfrac{1}{1:v - \cfrac{1}{3:v - \cfrac{1}{5:v - \cfrac{1}{7:v - 1 \& c.}}}} \tag{13.45}$$

Lambert then demonstrated that the right-hand side of the equation must be irrational if $v \neq 0$ is a rational number. But because $\tan(\pi/4) = 1$ is rational, it follows that $\pi/4$, and therefore also π itself, must be irrational.

After Lambert, Legendre proved in 1794, that π^2 is also irrational.

Arctan formulae

in 1706 John Machin (1680–1752) outdid Sharp's 71 decimal places with exactly 100 decimal places, all of them were correct. He was the first mathematician to use a combination of two arctan series in a single expression. His arctan formula became extremely popular:

$$\pi = 16 \arctan \frac{1}{5} - 4 \arctan \frac{1}{239} \tag{13.46}$$

Machin calculated each of the two arctan expressions using Gregory's series

$$\arctan \frac{1}{x} = \frac{1}{x} - \frac{1}{3x^3} + \frac{1}{5x^5} - \cdots \qquad (13.47)$$

For the first arctan expression with $x = 5$ he had to calculate around 70 terms in the series (13.47), but for the second expression with $x = 239$ he only needed 20 terms.

John Machin did not publish his famous achievement himself; his $(1 + 100)$-digit π and his arctan formula appeared in William Jones's previously mentioned work, the *Synopsis Palmariorum Mathesos*[68], in which the symbol π was also proposed for the first time. In this way, this work by a self-made mathematician contains two major innovations regarding π.

For more than 200 years afterwards no one came up with a better method of calculating π than to use combined arctan formulae. However, mathematicians soon learned how to construct other such relations after the model of Machin's formula which might offer scope for optimisation. C.F. Gauss (1777–1855) developed some especially effective variants. Thus it came about that the arctan method remained in use for more than 250 years until 1970, although of course this is far short of the life of Archimedes's method. Even the early computer calculations almost exclusively made use of arctan formulae, and one of those used most frequently was actually the good old Machin formula (5.20).

After Machin many other mathematicians took up the task, and every few years a new π world record would be set. But as the number of decimal places grew, so too did the risk of making a mistake. Thus we hear of several instances of calculation errors which in each case were revealed inevitably the next time the record was beaten. For example, although Fautet De Lagny (1660–1734) discovered 27 digits more than Machin in 1719, it turned out when the world record was next beaten by Vega (1754–1802), that his 112th decimal place was incorrect. However, as this was the only incorrect digit, it seems likely that it was a case of a transcription error rather than that the calculation itself was at fault. Vega himself also made a mistake – of the 143 digits he produced in 1789, only 126 were correct. Five years later Vega improved on his own record with a π that was 140 digits long, but the last four of these were also incorrect.

Even today, the last digits calculated are always the most suspect. The current world record holder Yasumasa Kanada cut off 200 digits from his 206.1 billion π digits to be on the safe side, although his test calculation only differed from the actual calculation in the last 45 digits. No doubt he thought there were enough digits anyway.

Euler too was a victim of De Lagny's error, as he reproduced De Lagny's number in his *Introductio*. Since then, publishers thinking of issuing a new edition of this important and interesting work have been faced with the dilemma as to whether they should simply correct the error or transfer it for the sake of historical truth. A recent English edition dating from 1988 chose the latter approach [52, p. 101], while an older German edition published in 1885 [53, p. 95] decided on the first approach.

In the 18th century, important mathematicians were also at work on π in Japan, although they remained behind their European counterparts. in 1722 Takebe Kenko (1664–1739) obtained 41 decimal places for π^2 using a series for π^2 (16.86), while in 1739 Matsunaga obtained 50 decimal places. Thereafter, according to Beckmann [18, p. 102], the Japanese seem to have had more sense than their European colleagues; they continued to study series yielding π, but wasted no more time on digit hunting. Today Beckmann would not be able to say this, as in recent years no one has pursued the hunt for π digits more vigorously than the Japanese Yasumasu Kanada and his team.

The Austrian mathematician Lutz von Strassnitzky (1803–1852) exploited an unusual opportunity which came his way. in 1840, the "famous mental computer" Zacharias Dase (1820–1861) visited him in Vienna and attended his lectures on elementary mathematics. As he was whiling away the time with the "most colossal, but pointless calculations", Strassnitzky persuaded him to do perform some research which "at least he would be able to use", namely the calculation of π to 200 decimal places. Among the formulae it was suggested he should use, Dase chose the following arctan formula which does not converge as well as Machin's formula but is well suited for paper and pencil (and rubber!) [112]:

$$\frac{1}{4}\pi = \arctan \frac{1}{2} + \arctan \frac{1}{5} + \arctan \frac{1}{8} \qquad (13.48)$$

In under two months Dase had finished. Dase's 200 decimal places were 8 fewer than William Rutherford had achieved twenty years earlier, but they were still a world record as only 152 of Rutherford's digits were actually correct.

Dase was known throughout Germany for his calculation skills. He was able, for example, to multiply two 8-digit numbers in his head in 54 seconds, while 200-digit numbers took him only 8 hours and 45 minutes, a feat which one can scarcely imagine. He offered to calculate some mathematical tables, so Gauss suggested he should expand the existing prime factorisation tables. Dase took up this suggestion and, with financial support from the Hamburg Academy of Sciences, he calculated the prime numbers in all the numbers between 7 and 9 million. Beckmann concludes his report on Zacharias Dase with the following commentary: "Evidently Carl Friedrich Gauss, an innovator in so many areas, was also the first to introduce payment for computer time" [18, p. 107].

The last records obtained using paper and pencil

The sad tale of William Shanks (1812–1882) took place against the backdrop of the mid-19th century. Shanks performed no fewer than four π calculations by hand. In 1851 he calculated first 315 and then 530 decimal places. Two years later he attained 607, and finally twenty years later as many as 707. In 1853 he published a complete book, which contained not only the 607 digits but a detailed description of how he had performed the 530 digit calculation. In the foreword he wrote:

"Towards the close of the year 1850, the Author first formed the design of rectifying the Circle to upwards of 300 places of decimals. He was fully aware, at that time, that the accomplishment of his purpose would add little or nothing to his fame as a Mathematician, though it might as a Computer; nor would it be productive of anything in the shape of pecuniary recompense at all adequate to the labour of such lengthy computations."

This was somewhat pessimistic, as Shanks's version of π was accorded the great honour of being exhibited at the big world exhibition in Paris in 1937 (see also page 50).

Shanks's calculations concealed a labour of Hercules. In the course of his calculations he had to determine some logarithms accurately up to 137 places and to calculate the exact value of 2^{721}. A Victorian commentator observed, "These huge chains of calculations not only prove the persistence and accuracy of this mathematician, but they are also a proof that computational skill and courage are increasing in society." [119, p. 51]

But fate had dealt Shanks an unlucky hand. To quote J.W. Wrench: "(His 607 decimal places were) incorrectly calculated beyond 527 decimal places. The accuracy was further vitiated by a blunder committed by Shanks in correcting his copy prior to publication, with the result that similar errors appear in decimal places 460–462 and 513–515. These errors persist in Shanks's first paper of 1873 containing the extension to 707 decimals of his earlier approximation. His second paper of that year, which contained his final approximation to π (of 707 decimals), gives corrections of these errors; however, there appears an inadvertent typographical error in the 326th decimal place of his final value. In retrospect, we now realize that Shanks's first value (530 decimal places) published in 1853 was the most accurate he ever published." [122]

The error in the 528th digit of Shanks's various π calculations remained *unproven* for many years. The person who got the furthest to checking it was Richter in 1854; however the latter did arrive at the wrong decimals. It was not until 1946, 90 years after Shanks, that the error was demonstrated by Daniel Ferguson, incidentally in what may have been the last π calculation to have been undertaken by hand.

However, Shanks's π had been *suspected* of being incorrect much earlier. The first person who became suspicious about the accuracy of Shanks's π was the mathematician Augustus de Morgan (1806–1871), a severe critic of the would-be circle squarer and inventor of the *morbus cyclometricus*, defined as the *circle-squaring disease*. He counted the frequency of every digit occurring in the 607 decimal version of Shanks's number and concluded from the deficiency of 7's that Shanks must have made a mistake. Unfortunately he was right.

This was all very sad as it only went to show how unfair the world can be: a person can sweat away at his labours for years only to have someone else glance at the results and show that they are wrong in a matter of an hour.

As the 19th century came towards its close, there was a major new breakthrough on π research. What had already been clear to mathematical scholars such as Michael Stifel (1487–1567) in the 16th century became certain fact in 1882, when it was established beyond doubt that π is not only irrational but also transcendental. Stated formally this means that π cannot be the root of any algebraic equation containing only rational coefficients. A corollary of this is that it is not possible to square the circle with a compass and straight edge. Proof of this was certainly in the air by 1873, when Charles Hermite (1822–1901) proved the transcendence of e. Decisive proof that π too is transcendental was provided a few years later, in 1882, by Ferdinand Lindemann (1852–1939). His rationale, which is very heavy going, was later simplified a number of times and today is available in an almost elementary form, e.g. see [49, pp. 103–109]. Like the proof of π's irrationality, the proof of its transcendence is once again indirect. The argument goes that if e^x is the root of a rational algebraic equation, then x must be transcendental (and vice versa). In fact, we know from Euler's famous formula (13.34) that $e^{i\pi}$ *is* a rational number, i.e. -1. Therefore $i\pi$ must be transcendental and hence π also.

The last world record calculation of π to have been performed by hand, or at most with the aid of an electromechanical hand-operated desk calculator, was performed in the years 1945-1949. Ferguson and John Wrench Jr. produced the last and best value yet, 1120 decimal places, before the advent of the computer era.

The computer era

It was at the end of the 1940s that computers entered the π arena, signalling an end to the era of only a few hundred π digits.

In 1949 one of the first electronic computers, the ENIAC (Electronic Numerical Integrator and Computer), was programmed to calculate 2,037 decimal places of π, using John Machin's arctan formula (5.20). The calculation took 70 hours to perform, i.e. more than 2 minutes per digit. In 1957, F.E. Felton attempted to calculate 10,000 decimal places, but due to a hardware fault only the first 7,480 places were correct. The goal of reaching 10,000 digits was achieved the following year by F. Genuys on an IBM 704 in 100 minutes [13].

On 29 July 1961, two IBM researchers, Daniel Shanks and John W. Wrench Jr., reached the 100,000 mark. They used an IBM 7090, which took 8 hours and 43 minutes to perform the calculation. They wrote a detailed account, still quite readable even today, of their project [106], in which they bring out the struggle they had with their algorithms. Like their predecessors they only had arctan series (one from Størmer (5.19) for the actual calculation and one from Gauss (5.22) to check it), but they came up with some interesting optimisation tricks in the course of their work.

Several other calculations increased the number of known π digits by 1973 to 1 million places, using only classical arctan formulae. The calculation of π to 1 million decimal places took just under one day on a CDC 7600.

Even though computers were becoming more powerful, the calculation of still more decimal places would have run into insurmountable obstacles at some point. Daniel Shanks, whom we mentioned above, still believed in 1961 that no one would ever succeed in calculating more than 1 million π digits. Castellanos [41] commented on this statement by quoting the Chinese saying that it is dangerous to make predictions especially when they refer to the future.

The Chinese saying did not stop Peter Borwein predicting in a lecture in 1995 that we would never discover the 10^{51}th digit of π. Even if 10^{51} is a very big number and Peter Borwein an extremely distinguished π researcher, this is a bold statement indeed.

13.4 High-performance algorithms

The third era in the hunt for π digits began in about 1980 when the world record still stood at around 1 million decimal places. It was around this time that the two breakthroughs occurred which together would enable the calculation of π digits to new orders of magnitude.

Fast multiplication

The first of these occurred in 1965 when the process of multiplying two (long) numbers was speeded up dramatically and to a certain extent out of proportion and the old "school multiplication" approach was superseded. Under the new method, known as *fast Fourier transform (FFT) multiplication*, the two multiplicands are first "Fourier"-transformed; they are then subjected to an pointwise multiplication process which can be executed in linear time, and finally the vector product is transformed back again. The critical factor in this multiplication process is the fact that the three Fourier transforms can be executed very quickly, i.e. at a rate better than quadratically (see pages 135ff.).

The discovery of this multiplication algorithm came more or less by chance:

"During a session of the scientific advisory committee of the American President [Lyndon B. Johnson in 1964] Richard W. Garwin established that John W. Tukey, who was one of the attendees, was trying to develop some programs for the Fourier transform[5]. Garwin ... asked ... and Tukey provided an outline of what was later to be the famous Cooley-Tukey algorithm."

It just goes to show how a President has his uses!

"Garwin went to the computer centre of the IBM research centre in Yorktown Heights to have the procedure programmed. James W. Cooley had not been working at the IBM research centre for long and, in his own words, was entrusted with the assignment because he was the only one who did not have anything important to do ..." [38, p. 8].

It was only a short while after publication of the Cooley-Tukey algorithm that the search for historical sources began. Someone who read their paper described an earlier program of his own with similar performance and attributed it to a method by G.C. Danielson and C. Lanczos dating from 1942. The authors for their part referenced essays by C. Runge and H. Koenig dating from the years 1903 and 1924 [44]. But these were not the earliest attempts at using FFT. It seems that once again it was C.F. Gauss who was the first to have had the all-important idea of speeding up the Fourier transform and in fact he did so 100 years (1805) before anyone else [65].

The time required for school multiplication is of quadratic order, whereas FFT multiplication, of which there are today many variants, only requires the order of $N \cdot \log_2 N$ (see page 140). With FFT multiplication, the time it takes to multiply n-digit multiplicands does not increase much more quickly than n itself.

π-specific algorithms

But as well as FFT multiplication, a second branch needed to be added to the tree of knowledge before today's π feats would become feasible, namely the discovery of high-performance algorithms developed specifically with the aim of calculating π more quickly and to additional decimal places. This was achieved from 1976, and as a result the arctan methods which had been virtually all that was available up to then were completely replaced.

One of the new methods was based on series which the Indian mathematician Srinivasa Ramanujan (1887–1920) had developed back

---

[5] Moreover, John Tukey is the inventor of the word "bit", short for binary digit, in 1946, and he also seems to be responsible for introducing the term "software" in the sense as it is used today, in 1958 ("The Economist" June 3rd 2000, p. 96)

in 1914. One of these is the following series, which is significantly faster, by around a factor of 5, than the best series known before then:

$$\frac{1}{\pi} = \frac{\sqrt{8}}{9801} \sum_{n=0}^{\infty} \frac{(4n)!}{(n!)^4} \cdot \frac{[1103 + 26390n]}{396^{4n}} \tag{13.49}$$

A second significant discovery in this area was the Gauss AGM algorithm. It consists of an iteration method such that, when executed, the number of correct π digits is doubled during each iteration stage, thus converging quadratically on π. In this way, for example, the first 9 iterations produce successively 1, 4, 9, 20, 42, 85, 173, 347 and 697 correct digits of π [13, p. 53].

This algorithm was independently discovered in 1976 by Eugene Salamin and Richard Brent. But the underlying formula had already been discovered 170 years earlier by the German mathematician Carl Friedrich Gauss (see Chapter 7). Salamin at least was aware of the origin of the algorithm when he published it.

In 1985, the Borwein brothers added to the stock of π-specific high-performance algorithms when they published iteration procedures in the style of the Gaussian AGM algorithm which converge at a rate better than quadratically. Thus, in the "quartic Borwein iteration" four times as many correct digits are produced during each iterative step so that in 1999 the present world record holder, Yasumasa Kanada, only needed twenty iterations to calculate 206.1 billion digits of π. The Borweins derived their method from the same theory of modular functions which had also led Ramanujan to his series.

It was thanks to these discoveries that, in an unprecedented boom spanning the 19 years from 1981 to 1999, the π world record was bettered a total of 26 times, often by more than double the previous record, from 2 million to currently 206.1 billion digits. Our table on page 206 lists the individual landmarks.

The new π digit hunters

It is striking that 18 out of the 39 world records are linked with the name Yasumasa Kanada (the stress falls on the second syllable). This man is Research Adviser in the Department of Science at Tokyo University. He heads a laboratory, *Kanada laboratory*, which specialises amongst other things in "high performance computing". To perform this work, Kanada has used special supercomputers, most recently a Hitachi SR8000 which contains 128 processors. His feats of calculation have been accomplished with these and ever more sophisticated implementations of the Borwein algorithms.

Unfortunately Kanada is not given to talking about his accomplishments. At the time of writing (April 2000) he has only just brought out a paper on his third most recent world record (51.5 billion digits) [113] (in Japanese) and he has issued only brief statements on the Internet regarding his latest records (68.7 and 206.1 billion digits). Kanada appears not to answer all the e-mails which are sent to him, so we know relatively little about the details of his work.

Even less has been divulged by Kanada's two "rivals", the Chudnovsky brothers, who have six entries in the list of world records. However, an excellent biographical report about the two brothers, *The Mountains of Pi*, by Richard Preston appeared in the *New Yorker* in March 1992 [92]. This essay describes the lonely life of David and Gregory Chudnovsky in their apartment in Manhattan, where they achieved their world records on a home computer called *m-zero* which they assembled from parts purchased in department stores. The money used to finance their work came from their wives, for the Chudnovskys have been unsuccessful at obtaining university posts although they have established track records in number theory. One reason for this, Preston remarks bitterly, is that Gregory suffers from a muscular disorder *Myasthenia gravis* which keeps him bed-ridden.

David Blatner, author of *The Joy of* π [28], an attractive little book which came out in 1997, apparently tried to track down the Chudnovsky brothers. He found out that they had recently opened a new office at the Brooklyn Polytechnic University, called the *Institute for Mathematics and Advanced Supercomputing*, which has a staff of precisely two (i.e. the Chudnovsky brothers themselves). They no longer live in Manhattan. Shortly before moving, they ran one more π calculation on their *m-zero*, this time to 8 billion digits, which, including checking, took up a week of computing time. This was a new record. They then gave up the chase to achieve further records and destroyed their home computer.

It seems that the experts do not trust the Chudnovskys entirely. When Yasumasa Kanada announced one of his world records, the Chudnovsky brothers said that they had already gone way past that figure. Whereupon Kanada sent them a few digits from "his" π, along with the position from where he had taken them. He asked the brothers to send him an analogous section from their π calculation for verification purposes. But the Chudnovsky brothers have never responded.

The only others to have held a world record since 1981 apart from Kanada or the Chudnovsky brothers are Gosper and Bailey.

As mentioned above, Gosper calculated 17 million decimal places in 1985, using the Ramanujan series (13.49). His computer was not a supercomputer but a comparatively simple Symbolics 3670 workstation which, however, compared with conventional computers, has the advantage of unlimited arithmetical precision. When announcing his record, Gosper stressed the fact that he had not calculated simple decimal places but more sophisticated continued fraction places. His sequence thus did not begin with 3,1,4,1,5, but with 3,7,15,1,292. Gosper hoped to attain a deeper understanding of the nature of the number π by adopting this representation form, since for number theoreticians a continued fraction is definitely a more informative alternative. Sadly the continued fraction did not reward the effort of his calculations but only suggested "normal probability" (see also Chapter o4 on this point). Nevertheless, the calculation did confirm that the continued fraction of π does not follow any model, which is actually a useful piece of information, since, for example, the numbers e, $\sqrt{2}$ or $\sqrt{3}$ do follow a model.

To crown it all, Gosper ended up having to convert his continued fraction to the decimal form solely in order to be able to check it.

David Bailey is a scientist at the NASA Ames Research Center in California, USA. He has published extensively on high-performance calculations and is one of the authors of the BBP algorithm. Between 7 and 9 January 1986 in a total of 28 hours he calculated π to 29 million decimal places and wrote a fine essay about it [9]. In the essay he states amongst other things that the main purpose of his calculation was to test the hardware (of a Cray-2 supercomputer), the operating system and the compiler. He wrote his calculation program in FORTRAN because the compiler for that programming language was able to exploit the vector processor properties of the Cray particularly well. The actual calculation required 12 iterations of the quartic Borwein algorithm and checking it required 24 iterations of the quadratic Borwein algorithm. The report also contains statistics on the frequencies of different numbers, but these do not produce any striking results.

13.5 The hunt for single π digits

We have already quoted the observation by Adamchik and Wagon that the "2000-year search [for ever more π digits is changing] direction". Today more and more is heard about a new type of π record, which was proclaimed with the words "The forty trillionth bit of π is 0".

The background to this is the fascinating "BBP algorithm", which is named after the initial letters of its authors, David Bailey, Peter Borwein and Simon Plouffe, and was only published recently, in September 1995. With this algorithm it is possible to calculate individual hexadecimal positions *in the middle* of π without having to calculate all the preceding digits. This method is described in detail in Chapter 10.

This algorithm has opened up a new stage in the quest for π digits, namely the search for digits as far removed from the beginning of π as possible. The first landmark was set by the three researchers themselves with the 10 billionth hexadecimal π digit, corresponding to the 12 billionth decimal place. At the time of publication of their data (September 1995), this digit was almost twice as far into π as the last known digit obtained by the conventional route.

Within only a year later a student called Fabrice Bellard[6] had caught up. He discovered an even better formula and used it to calculate the 100 billionth (October 1996) and the 250 billionth (September 1997) hexadecimal digits of π, which were a factor of 10 and 25, respectively, more distant. But even these achievements were overtaken almost immediately by the disclosure of some hexadecimal digits in the trillionth area such as the 1.25 trillionth (August 1998), the 10 trillionth (February 1999) and the 250 trillionth (11 September 2000)[7], which in turn bettered the previous records by factors, respectively, of 5, 20 and 1000. The holder of both records is 19-year-old Colin Percival, a student at the Simon Fraser University in Burnaby, Canada, the same establishment where the Borwein brothers and Simon Plouffe work. No doubt we will hear many more great things about this young man. He began his university studies in mathematics at the age of 13 in parallel to his secondary school education, and has successfully completed an unusually major project while still in his teens.

Although the formula Percival used was not new (in fact, it was the Bellard formula, 10.8), for the calculation itself he used an unusual

---

[6] http://www-stud.enst.fr/~bellard/
[7] http://www.cecm.sfu.ca/projects/pihex/announce1q.html

approach which may well become a standard approach in years to come (see also Chapter 15). What was new about it was that he obtained the co-operation of 1734 Internet computers based in 56 different countries which all calculated parts of his project in their "dead time". Percival then put the results together. In this way he obtained the benefit of almost 700 years of computing time in an elapsed period of only two years.

No doubt he will produce another, even more spectacular result soon.

With this we end our excursion through history and through 4000 years of research into π. The most important milestones are summarised one more time in the tables on the following pages. What we know now is the beginning and the middle. The end is not yet in sight.

13.5 The hunt for single π digits

Who?	When?	No. of decimals	Algorithm[1]	Comment
Babylonians	2000? BC	1	$3 + 1/8$	
Egyptians	2000? BC	1	$4 \cdot (8/9)^2$	
Indians	600? BC	0	$4 \cdot (9785/11136)^2$	
Bible	440? BC	0	$\pi = 3$	
Plato	c. 380 BC	2	$\sqrt{2} + \sqrt{3}$	
Archimedes	c. 250 BC	2	pgm(96)	$\frac{223}{71} < \pi < \frac{22}{7}$
Zhang Heng	c. 130 AD	1	$\sqrt{10}$	
Ptolemy	150	3	377/120	
Wang Fan	c. 250	1	142/45	
Liu Hui	263	5	pgm(192)	
Tsu Chhung-Chih	c. 480	7	pgm(12288)	
		6		$\pi = \frac{355}{113}$
Aryabhata	499	3	pg(384)	
Brahmagupta	640?	1	$\sqrt{10}$	
Alkarism	830	3	62832/20000	
Fibonacci	1220	3	864/275	
Dante	c. 1320	3	$3 + \sqrt{2}/10$	
Al-Khashî	1430	16	pgm($3 \cdot 2^{28}$)	2π
Nilakantha	c. 1501	9	104348/33215	
Viète	1579	9	pgm($3 \cdot 2^{17}$)	
Romanus	1593	15	pgm(2^{20})	
Ludolph Van Ceulen	1596	20	pgm($15 \cdot 2^{35}$)	
Ludolph Van Ceulen	1615	35	pgm(2^{62})	
Snell	1621	34	pgm(2^{30})	
Grienberger	1630	39	pgm()	last use of p'gons
Newton	1665	15	N	first use of series
Sharp	1699	71	G	
Machin	1706	100	M	
De Lagny	1719	111	G	127 total
Matsunaga	1739	50	H	
Vega	1794	136	E	140 total
Rutherford	1824	152	E	208 total
Dase	1844	200	S	
Clausen	1847	248	M	
Lehmann	1853	261	H	
W. Shanks	1853	530	M	
Ferguson	1945	530	L	
Ferguson	1946	620	L	last hand calc'ion
Ferguson	1947.01	710	L	desk calculator
Ferguson, Wrench	1948.01	808	M	
Smith, Wrench	1949.06	1,120	M	

[1] pgm(x) = Polygons with x sides.
Series: E=Euler (5.25), G=Gregory (5.3), H=Hutton (5.17), L=Loney (5.18), M=Machin (5.20), N=Newton (16.63), S=Strassnitzky (16.108),

Table 13.1. History of π in the pre-computer era

Who?	When ?	No. of decimals	Alg[1]	Time	Computer
Reitwiesner	1949.09	2,037	M	70 h	ENIAC
Nicholson, Jeenel	1954.11	3,092	M	0:13 h	NORC
Felton	1957	7,480	K	33 h	Pegasus
Genuys	1958.01	10,000	M	1:40 h	IBM 704
Felton	1958.05	10,020	K	33 h	Pegasus
Guilloud	1959	16,167	M	4:18 h	IBM 704
Shanks, Wrench	1961.07	100,265	S	8:43 h	IBM 7090
Guilloud, Filliatre	1966.02	250,000	G	41:55 h	IBM 7030
Guilloud, Dichampt	1967.02	500,000	G	28:10 h	CDC 6600
Guilloud, Boyer	1973	1,001,250	G	23:18 h	
Miyoshi, Kanada	1981	2,000,036	K	137:18 h	FACOM
Guilloud	1981-82	2,000,050			
Tamura	1982	2,097,144	G2	7:14 h	MELCOLM
Tamura, Kanada	1982	4,194,288	G2	2:21 h	HIT M-280H
Tamura, Kanada	1982	8,388,576	G2	6:52 h	HIT M-280H
Kanada, Yoshino, Tamura	1982	16,777,206	G2	< 30 h	HIT M-280H
Gosper	1985.10	17,526,200	R		Symbolics 3670
Bailey	1986.01	29,360,111	B4	28 h	CRAY-2
Kanada, Tamura	1986.09	33,554,414	G2	6:36 h	HIT S-810/20
Kanada, Tamura	1986.10	67,108,839	G2	23 h	HIT S-810/20
Kanada et al.	1987.01	134,217,700	G2	35:15 h	NEC SX-2
Kanada, Tamura	1988.01	201,326,551	G2	5:57 h	S-820
Chudnovsky's	1989.05	480,000,000 ?	C	≈ 6 mon	CRAY-2
Chudnovsky's	1989.06	525,229,270	C	> 1 mon?	IBM-3090
Kanada, Tamura	1989.07	536,870,898	G2	67:13 h	HIT S-820/80
Chudnovsky's	1989.08	1,011,196,691?	C	> 2 mon ?	IBM-3090
Kanada, Tamura	1989.11	1,073,741,799	G2	74:30 h	HIT S-820/80
Chudnovsky's	1991.08	2,260,000,000?	C	250 h ?	m-zero
Chudnovsky's	1994.05	4,044,000,000?	C	unknown	unknown
Takahashi, Kanada	1995.06	3,221,225,466	B4	36:52 h	HIT S-3800
Takahashi, Kanada	1995.08	4,294,967,286	B4	113:41 h	HIT S-3800
Takahashi, Kanada	1995.10	6,442,450,938	B4	116:38 h	HIT S-3800
Chudnovsky's	1996.03	8,000,000,000	C	unknown	1 week ?
Takahashi, Kanada	1997.04	17,179,869,142	G2	5:11 h	HIT SR2201
Takahashi, Kanada	1997.05	34,359,738,327	B4	15:19 h	HIT SR2201
Takahashi, Kanada	1997.07	51,539,607,510	B4	29:03 h	HIT SR2201
Takahashi, Kanada	1999.04	68,719,470,000	B4	32:54 h	HIT SR8000
Takahashi, Kanada	1999.09	206,158,430,000	G2	37:21 h	HIT SR8000

[1] Series: M=Machin (5.20), G=Gauss (5.22), K=Klingenstierna (5.21), S=Størmer (5.19), R=Ramamanujan (13.49), C=Chudnovsky (8.7). Iterations: G2=Gauss AGM (alg 7.5), B4=Borwein quartic (alg 9.2)

Table 13.2. History of π in the computer era [113]

13.5 The hunt for single π digits

Who?	When ?	Hexadecimal Position	Formula	Hexadecimal Sequence
Bailey, P. Borwein, Plouffe	1995.09	10^{10}	BBP (10.1)	921C73C683 ...
Bellard	1996.10	10^{11}	Bellard (10.8)	9C381872D2 ...
Bellard	1997.09	$2.5 \cdot 10^{11}$	Bellard (10.8)	87F72B1DC9 ...
Percival	1998.08	$1.25 \cdot 10^{12}$	Bellard (10.8)	07E45733CC ...
Percival	1999.02	$10 \cdot 10^{12}$	Bellard (10.8)	A0F9FF371D ...
Percival	2000.09	$250 \cdot 10^{12}$	Bellard (10.8)	E6216B069C ...

Table 13.3. History of digit extraction records

14. Historical Notes

14.1 The earliest squaring the circle in history?

On the circumference of the circle of radius 5, there are 12 points whose co-ordinates are integers. These are the 4 points at which the circle is intersected by the co-ordinate axes ($\pm 5, 0$) and ($0, \pm 5$) and the 8 angular vertices of the right-angled triangles whose non-hypotenuse sides are 3 and 4, i.e. points ($\pm 4, \pm 3$) and ($\pm 3, \pm 4$). This is illustrated in the left-hand diagram, which shows the first quadrant of the circle:

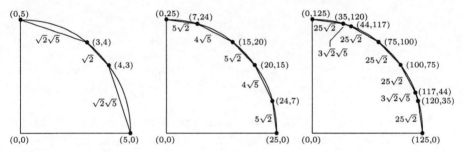

The 12 points define an (irregular) 12-sided polygon, 8 of whose sides are of length $\sqrt{2} \cdot \sqrt{5}$, and the other 4 are of length $\sqrt{2}$. Its diameter is 10 and its perimeter $8 \times \sqrt{2}\sqrt{5} + 4 \times \sqrt{2} = 30.9550\ldots$; both together produce an approximation for π of $3.09550\ldots$.

If the circle is enlarged by a factor of 5 to radius 25 (centre diagram), we find the original 12 points (or rather, their equivalents) ($\pm 25, 0$), ($\pm 20, \pm 15$), ($\pm 15, \pm 20$) and ($0, \pm 25$) once again, but this time there are another 8 extra points which have integer co-ordinates ($\pm 24, \pm 7$) and ($\pm 7, \pm 24$). The side lengths of the irregular 20-sided polygon thus defined add up to $12 \times 5\sqrt{2}$ plus $8 \times 4\sqrt{5}$, and they produce the approximation $\pi \approx 3.12814\ldots$.

If the circle is magnified a second time by a factor of 5, we obtain 8 further points which have integer co-ordinates, ($\pm 117, \pm 44$) and ($\pm 44, \pm 117$) (diagram on the right). The resulting 28-sided polygon with diameter 250 has 20 sides of length $25\sqrt{2}$ and 8 sides of length $3\sqrt{2}\sqrt{5}$, and the approximation is now 3.13200....

It will be seen that every time the radius is increased by a factor of 5, 8 new vertices occur whose co-ordinates are integers[1]; even more important for calculation purposes is the fact that all the side lengths of the resulting 36-, 44- etc. sided polygons can be expressed solely using square roots of 2 and/or 5. This captivatingly simple procedure, which would doubtless have thrilled Pythagoras, was discovered by Franz Gnaedinger of Zurich [59]; the complete proof was contributed by Christoph Poeppe. In arriving at this discovery, Gnädinger identified additional simple geometric constructions for the formation of the vertex co-ordinates and also devised simple methods to approximate the square roots from 2 and 5 to any degree of accuracy. For the roots from 2, for example, he specifies the following number stack, whose construction rule is easy to guess:

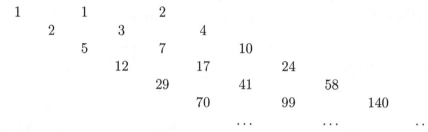

The fractions 10/7 and 7/5 are simple approximations for $\sqrt{2}$. 24/17 and 17/12 are better approximations, and 140/99 and 99/70 are already very good approximations for $\sqrt{2}$.

Now Franz Gnädinger is first and foremost an Egyptologist, and so he looked to see whether the good old Egyptians (who of course were very young in the pharaonic era, due to their low life expectancy, and looked at the world with just the same curiosity that young people do today) had known of this method. His conclusion is that they probably did.

To start with, the Egyptians had a motive for measuring circles. The hieroglyph for the sun god Re was a circle. If one could only understand the circle and fathom its secret number, one would be able to share in the power of Re.

---

[1] The co-ordinates of the new points are generated iteratively as follows:
$x_0 := 0, \quad y_0 := 1$ Thereupon for $k = 0, 1, 2, \ldots$:
$x_{2k+1} = 3y_{2k} - 4x_{2k}, \quad y_{2k+1} = 4y_{2k} + 3x_{2k}$
$x_{2k+2} = 4y_{2k+1} - 3x_{2k+1}, \quad y_{2k+2} = 3y_{2k+1} + 4x_{2k+1}$

But the Egyptians were also quite precocious in mathematical matters. They already knew about Pythagorean triangles (right-angled triangles whose side lengths are integers), the simplest of which have side lengths which are multiples of 3, 4 and 5 and were referred to by the Egyptians as "holy triangles".

According to Gnädinger, all the elements of a method for approximating π and even the most important numbers are to be found in King Zoser's funerary complex at Saqqara, in the Cheops pyramid and the Chephren pyramid, for example:

- Saqqara: 17/12 for root 2
- Saqqara: numerous Pythagorean triads
- Chephren pyramid: half cross-section = holy triangle
- Cheops pyramid: holy triangle 15-20-25 in the royal chamber
- Cheops pyramid: 140/99 for root 2 and 161/72 for root 5
- Cheops pyramid: base×2:height = 22:7 or approximately π

Gnädinger is especially interested in the visionary architectural genius Imhotep, who c. 2600 BC designed the monumental and graceful step pyramid at Saqqara near the former Memphis, to the fame and eternal life of Pharaoh Zoser. Gnädinger cites other Egyptologists in support of his view that this Imhotep could well have developed extremely good approximations for π based on the above procedure and the rational approximations for the roots of 2 and 5 which have been handed down. After just two iterations, with irregular 20-sided polygons, he would have arrived in systematic manner at the approximation $\pi \approx 3.128\ldots$, which is significantly better than the value of $(\frac{16}{9})^2 = 3.160\ldots$ which is found in the Rhind Papyrus of 1850 BC and was probably discovered empirically (see page 167). This more accurate version of π would have been more than 700 years older.

14.2 A π law

100 years ago a law prescribing the value 3.2 for π was very nearly passed in the state of Indiana, USA. The story is no doubt mainly amusing to Americans, perhaps because the event took place on American soil and in any case everything turned out well in the end, or perhaps another reason why they like it could be that a former, less popular vice-president came from Indiana.

At the end of the nineteenth century, a country doctor named Dr. Edward Johnston Goodwin (1828?–1902) lived in the town of Solitude, Indiana. According to his biography, this man discovered the true value of the number π by supernatural means in the first weeks of March 1888.

Strictly speaking, the one revelation must have included several values of π, since, in a reader's letter to the fledgling "American Mathematical Monthly"[2], he actually gave 5 different values of π, from 2.56 to 4.0.

He applied for copyright for his discovery in several states, including Germany and Austria. He then persuaded a representative of his home town to introduce a Bill on his new mathematical discovery.

The proposed Bill was written by Dr. Goodwin himself. In this Bill he described π in words and – as he had previously done in the American Mathematical Monthly – in several ways. House Bill No. 246 contains, for example, the passage:

> ...disclosing the fourth important fact, that the ratio of the diameter and circumference is as five-fourth to four...

This "fact" results in $\pi = 16/5 = 3.2$. Elsewhere in the text three other values of π are given, namely, the argument that the area = (circumference/4)^2 produces $\pi = 4$, the argument that the area = $6 \cdot$ (circumference/4)^2/5 results in $\pi = 10/3 = 3.33333$, while $\pi = 32/9 = 3.555556$ is obtained from diameter$^2 = 9^2$ and circumference/4 = 8.

The Bill met with its first hurdle in the Indiana house of representatives on 5 February 1897. After deliberation in the responsible committee, the Bill was passed to the senate for confirmation, with the recommendation that it should be become law.

In defence of the representatives it must be said that their primary consideration was the commercial good of their state and in mathematical affairs they relied entirely to the reputation of Dr. Goodwin. One representative argued that if this Bill were passed, then the author would give his home state the right to use his discovery free of charge and, for example, to publish it gratis in school textbooks whereas everyone else would have to pay a royalty to use it.

Luckily for the reputation of Indiana Professor C.A. Waldo of Purdue University happened to be visiting the house of representatives that day and listened in on the committee session. He pointed out

---
[2] The magazine had actually existed earlier but under the name of "Ladies Diary".

the shortcomings of the Bill to the senators, who placed all further readings of the Bill on hold, and the matter was permanently laid to rest.

One of the historians recounting this event, David Singmaster, felt that Goodwin was due particular credit as one of only very few "circle squarers" to have possessed several values for π [110]. The mean value of all the π's which Goodwin served up in articles, interviews and in the above draft Bill, was calculated by Singmaster at 3.28907.

This story even appeared in the German newspaper "Die ZEIT" (1997/28, 4 July 1997). There it appeared with the following comment: "Before we laugh too loudly at the legislature of Indiana or at the general ignorance which prevailed in 1897, we should pause a moment and reflect on what fate the draft Bill would have met, had it been the subject of a referendum today."

14.3 The Bieberbach story

During the Nazi era Hitler's Germany formed the backdrop to a disgraceful episode concerning π. Between 1909 and 1933 the Jewish mathematician Edmund Landau (1877–1938) pursued a distinguished teaching career at the University of Göttingen, a renowned centre of excellence in mathematics whose prominent figures included, for example, David Hilbert, Felix Klein and Emmy Noether. In his lectures on differential and integral calculus he used the definition which today is regarded as standard that $\pi/2$ is the smallest positive zero of $\cos x$. In the telegram style which was characteristic of his writings, Landau wrote in a textbook, "The universal constant from theorem 262 will always be denoted by π [51, p. 130]."

In 1933 Landau was dismissed from his chair on grounds of his race. An important colleague of Landau's subsequently added insult to injury by gratuitously suggesting that Landau's dismissal was justified on mathematical/psychological grounds.

This Berlin colleague of Landau's was the function theoretician Ludwig Bieberbach. In June 1934 he wrote the following lines in a treatise on *Personality structure and mathematical creativity*:

"In this way ... the ultimate reason behind the courageous rejection which the students at Göttingen University meted out to a great mathematician, Edmund Landau, was that his un-German style in research and teaching had become intolerable to German sensitivities.

A people which has seen how alien desires for dominion are gnawing at its identity, how enemies of the people are working to impose their alien ways on it, must reject teachers of a type alien to it."

The English mathematician Godfrey H. Hardy, whom we encountered earlier as S. Ramanujan's patron (see page 106), responded to Bieberbach in these words [109]:

"There are many of us, many Englishmen and many Germans, who said things during the (First) War which we scarcely meant and are sorry to remember now. Anxiety for one's own position, dread of falling behind the rising torrent of folly, determination at all costs not to be outdone, may be natural if not particularly heroic excuses. Prof. Bieberbach's reputation excludes such explanations of his utterances; and I find myself driven to the more uncharitable conclusion that he really believes them true."

15. The Future: π Calculations on the Internet

In this chapter we introduce you to a procedure which enables π to be computed using a distributed technique and at the same time allows the results thus computed to be recycled for later computations. This approach means that large parts of the computational work can be shared out over many small computers on the Internet. This has the major advantage that ultra-expensive supercomputers only have to be used for those parts of the calculation which only they are capable of performing.

The underlying algorithm is called the *binary splitting algorithm*, or *binsplit algorithm* for short. This algorithm, which has been around for some time, was recently re-examined in more detail by Bruno Haible [62]. It is precisely the π series which have long since ceased to be used in the quest for world records that are now returning to favour. The critical factor here is, however, rapid convergence. With this in mind it is not surprising that the previously mentioned series developed by the Chudnovsky brothers (8.7), which converges with 15 decimal places per term, is now *the* candidate for the binsplit algorithm.

15.1 The binsplit algorithm

Just for once in this book, we invite you to attempt the task of arriving at a correct derivation yourself. But don't worry, this derivation is very short, very elementary and once you have done it you will be that much wiser.

To recap, the time taken to naively total a sum $\sum_{k=0}^{N-1} a_k$ is proportional to N^2. To improve it, one calculates the ratio r_k of successive summands:

$$r_k := \frac{a_k}{a_{k-1}} \tag{15.1}$$

(To avoid treating $k = 0$ as a separate case, one sets $a_{-1} := 1$.) Hence,

$$\sum_{k=0}^{N-1} a_k =: r_0\left(1 + r_1\left(1 + r_2\left(1 + r_3\left(1 + \ldots \left(1 + r_{N-1}\right)\ldots\right)\right)\right)\right) \quad (15.2)$$

Now we define

$$r_{m,n} := r_m\left(1 + r_{m+1}\left(\ldots(1 + r_n)\ldots\right)\right) \quad \text{where} \quad m < n \quad (15.3)$$

$$r_{m,m} := r_m \quad (15.4)$$

Then,

$$r_{m,n} = \frac{1}{a_{m-1}} \sum_{k=m}^{n} a_k \quad (15.5)$$

(No need to believe it, just work it out!)
In particular,

$$r_{0,n} = \sum_{k=0}^{n} a_k \quad (15.6)$$

Moreover, the following also holds:

$$r_{m,n} = r_m + r_m \cdot r_{m+1} + r_m \cdot r_{m+1} \cdot r_{m+2} + \ldots \quad (15.7)$$

$$\ldots + r_m \cdot \ldots \cdot r_x + r_m \cdot \ldots \cdot r_x \cdot [r_{x+1} + \ldots + r_{x+1} \cdot \ldots \cdot r_n]$$

$$= r_{m,x} + \prod_{k=m}^{x} r_k \cdot r_{x+1,n} \quad (15.8)$$

hence

$$r_{m,n} = r_{m,x} + \frac{a_x}{a_{m-1}} \cdot r_{x+1,n} \quad (15.9)$$

(where $m \leq x < n$). All the terms listed here are *rational* numbers, and therefore consist of *two integers*.

In this way we can formulate the binsplit algorithm by simply specifying a recursive binsplit function:

```
rational function r (rational function a, int m, int n)
{
    rational ret;

    if m==n then
    {
        ret := a(m)/a(m-1)
    }
    else
    {
        x := floor( (m+n)/2 )
```

```
        ret := r(a,m,x) + a(x) / a(m-1) * r(a,x+1,n)
    }
    return ret
}
```

Here, a(k) is a function which returns the kth term of the series to be summed. As an example let us take arctan(1/10):

```
function a(int k)
{
    if k<0 then   return 1
    else          return (-1)^k/((2*k+1)*10^(2*k+1))
}
```

The call r(a,0,N) returns $\sum_{k=0}^{N} a_k$.

If one looks at the terms which occur on the call r(.,0,7), it will be apparent that this procedure is better than the naive route. Here first of all is the recursive structure:

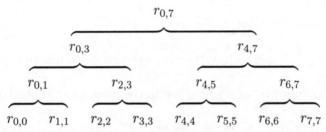

Every node corresponds to a call to the function r, the top one to the user call r(.,0,7,); all the lower nodes are recursive calls of r by itself.

Specifically, for our arctan(1/10) example we obtain

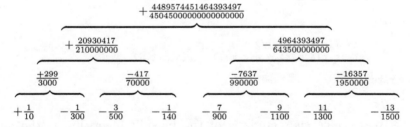

The numerators and denominators in row k (counted from the top) are approximately $N \log(N)/2^k$ digits long, where N is the number of digits in the end result. If we make the simplifying assumption that the time required to perform a multiplication operation increases linearly with N^1, it can be shown ([62]), that the total time required under

---

[1] Instead of the realistic $N \cdot \log(N) \log(\log(N))$, which can be achieved with the best known multiplication method, namely the Schönhage-Strassen method [104].

reasonable assumptions is proportional to $M(N) \cdot (\log(N))^2$, where $M(N)$ indicates the time taken for a multiplication operation of length N (In row k of the tree there are proportionally 2^k multiplication operations of length $N \log(N)/2^k$.) In this way, series with linear but sufficient convergence can be considered once more as candidates for achieving π world records.

We have actually cheated a little: it is true that the terms r(m,n) which occur in our tree are of a length proportional to n-m, but the a(n) which occur during recursion are proportionally n long, so that on the right-hand side of the tree they are always of maximal length. This is not actually so bad as the *quotients* a(k)/a(j) which occur are again only proportionally k-j long. We therefore specify a function which calculates this quotient without a detour involving relatively long terms:

```
rational function a(int k, int j)
{
    if j<0 then    return (-1)^k/((2*k+1)*10^(2*k+1))
        else       return (-1)^(k-j)*(2*j+1)/(2*k+1)/10^(2*(k-j))
}
```

The binsplit function which is effectively doing the calculation thus goes:

```
function r(function a, int m, int n)
{
    rational ret;

    if m==n then
    {
        ret := a(m,m-1)
    }
    else
    {
        x := floor( (m+n)/2 )
        ret := r(a,m,x) + a(x,m-1) * r(a,x+1,n)
    }
    return ret
}
```

A really nice characteristic of the binsplit algorithm is the fact that each fraction, once calculated, can be reused for later, expanded calculations. Let us assume that in the latest (N-long) world record r(.,0,N-1) has already been calculated. Let us suppose that technical progress means that we now have a computer which for the first time can calculate numbers $2N$ digits long. We therefore want to work out r(.,0,2N-1) for a new world record. This is not a problem, since according to formula (15.9) the following applies (with $x = N - 1$):

$$r_{0,2N-1} = r_{0,N-1} + a_{N-1} \cdot r_{N,2N-1} \tag{15.10}$$

We already have $r_{0,N-1}$, so what we need to compute is $r_{N,2N-1}$. However, this can actually be done on the "old" computer and could therefore also be calculated in advance. The new supernumbercruncher would then actually only need to perform the last but one iteration.

15.2 The π project on the Internet

Armed with the binsplit algorithm, we would now like to describe a project for calculating π on the Internet. Instead of numbers of digits, what we are concerned with here is memory size (there is only one proportionality constant throughout, typically, for example, 2 [decimal places per byte]). We will assume that our supercomputer has 1 terabyte (TB) of memory, i.e. 2^{40} bytes. This amount of memory can be halved 40 times until we are down to one byte; thus the binsplit recursion depth is 40 (in reality, a little smaller if the series converges rapidly). We now divide the work into 3 layers:

- Right at the top is a supercomputer. This will take over as few rows as possible (in the tree structure), ideally only the ones which really cannot be handled by any other computer.
- For the lower levels, we need as many voluntary Internet participants as possible (for example, you!) who have computer time available on their workstations (which must have a reasonable amount of memory). They will take over as many rows as possible. But as the amount of memory available on such PCs is significantly lower than 1 TB, we plan to increase the amount of virtual memory by a factor of 4 to 16 using "mass storage FFTs"[2]. A typical PC has around 64 MB, which mass storage FFTs can then effectively transform into 256 MB ($= 2^{28}$ bytes). This means that the Internet volunteers can each take over 28 rows, i.e. almost two-thirds of the necessary CPU work.

There is no doubt that the existing CPU resources are *easily* sufficient for this: CPU cycles equivalent to 500,000 years or *a whole geological era* of CPU time, capable of executing a million instructions per second, were recently applied on a distributed project (an

---

[2] These are fast Fourier transform (FFT) algorithms which can handle data volumes larger than the RAM storage available; they work with files on the hard disk.

attempt to break an encryption code). Ideally one would only use periods during which the computers are not doing anything for this project. There is one remaining problem: since for reasons of economy we do not want to burden the mainframe with all 12 remaining rows, there is a yawning gap.

- Hence, a company which manufactures really nice computers is making available to us enough computing time on a computer fitted with 16 GB of RAM (and a vast amount of hard disk storage space), on which we will be able to calculate the rows up to and including 128 gigabytes[3]. In this way the supercomputer will only need to process the rows with operations requiring 256, 512 and 1024 GB, a total of 3 out of 40.

Meanwhile we must not forget that the Chudnovsky series also requires a square root to be calculated. But this is not a major problem: for roots of integers, any number of good rational approximations can be specified. If $\frac{u_0}{v_0}$ is used as a starting approximation for $\sqrt{d}$, then the iteration produces

$$\frac{u_{n+1}}{v_{n+1}} := \frac{u_n^2 + d\,v_n^2}{2\,u_n\,v_n} \tag{15.11}$$

an improved approximation with twice the precision (see section 11.5).

If one really wants to transform the huge fraction into the digit sequence 3.14159..., another division operation is necessary, as the binsplit algorithm produces the numerators and denominators separately. This division entails 4 multiplication operations (see section 11.4).

The question of how to transfer the data thus generated requires some thought. Transfer of the (spatially distributed) lower rows to the middle rows will certainly take place over the Internet, in chunks of several hundred megabytes. It is therefore essential to have a good Internet connection. The chunks will add up to 2 TB in all, 1 TB for the numerators and 1 TB for the denominators. The transfer to the supercomputer will probably be performed by post, since at the end of the day it is several orders of magnitude cheaper to send a few kilograms of storage media than to transfer 2 TB over the Internet. Here the binsplit algorithm will demonstrate its compatibility with truly traditional methods of data transfer.

---

[3] Here too mass storage FFT will be used.

The project is feasible. We have analysed it and it looks good. It would be the start of a new era in π world records.

16. π Formula Collection

$$\pi = \frac{\text{circumference}}{\text{diameter}} \qquad (16.1)$$

$$\pi = \frac{\text{area of circle}}{\text{radius of circle}^2} \qquad (16.2)$$

$$\pi = 4\arctan 1 \qquad (16.3)$$

$$\pi = \arctan 1 + \arctan 2 + \arctan 3 \qquad (16.4)$$

$$\pi = -i\ln(-1) \qquad (16.5)$$

Nilakantha series, 15th century

$$\pi = \sqrt{12}\left(1 - \frac{1}{3 \cdot 3^1} + \frac{1}{5 \cdot 3^2} - \frac{1}{7 \cdot 3^3} + \frac{1}{9 \cdot 3^4} - \cdots\right) \qquad (16.6)$$

Nilakantha, 15th century [89]

$$\frac{\pi}{4} \approx 1 - \frac{1}{3} + \frac{1}{5} - \frac{1}{7} + \cdots \mp \frac{1}{p-1} \pm \frac{p/2}{p^2+1} \qquad (16.7)$$

where p is the last odd denominator $+1$

Nilakantha, 15th century [89]

$$\frac{\pi}{4} \approx 1 - \frac{1}{3} + \frac{1}{5} - \frac{1}{7} + \cdots \mp \frac{1}{p-1} \pm \frac{\frac{p^2}{4}+1}{\frac{p}{2}(p^2+4p+1)} \qquad (16.8)$$

where p is the last odd denominator $+1$

Nilakantha, 15th century [89]

$$\frac{\pi}{16} = \frac{1}{1^5+4\cdot 1} - \frac{1}{3^5+4\cdot 3} + \frac{1}{5^5+4\cdot 5} - \frac{1}{7^5+4\cdot 7} + \cdots \qquad (16.9)$$

Nilakantha, 15th century [89]

$$\pi = 3 + \frac{4}{3^3-3} - \frac{4}{5^3-5} + \frac{4}{7^3-7} - \frac{4}{9^3-9} + \cdots \qquad (16.10)$$

Nilakantha, 15th century [89]

$$\pi \approx 2 + \frac{4}{2^2-1} - \frac{4}{4^2-1} + \frac{4}{6^2-1} - \cdots \mp \frac{4}{p^2-1} \pm \frac{4}{2(p+1)^2+4} \qquad (16.11)$$

where p is the last even perfect square in the series

Nilakantha, 15th century [89]

$$\frac{\pi}{8} = \frac{1}{2^2-1} + \frac{1}{6^2-1} + \frac{1}{10^2-1} + \frac{1}{14^2-1} + \frac{1}{18^2-1} + \cdots \qquad (16.12)$$

Nilakantha, 15th century [89]

$$\frac{\pi}{8} = \frac{1}{2} - \frac{1}{4^2-1} - \frac{1}{8^2-1} - \frac{1}{12^2-1} - \frac{1}{16^2-1} - \cdots \qquad (16.13)$$

Nilakantha, 15th century [89]

Formulae of Leonhard Euler (1707–1783)

$$e^{i\pi} = -1 \text{ or: } \pi = -i\ln(-1) \text{ or: } e^\pi = i^{-2i} \qquad (16.14)$$

Euler, 1743

$$\arctan x = \frac{x}{1+x^2}\left(1 + \frac{2}{3}\left(\frac{x^2}{1+x^2}\right) + \frac{2\cdot 4}{3\cdot 5}\left(\frac{x^2}{1+x^2}\right)^2 + \cdots\right) \qquad (16.15)$$

Euler, 1755

$$\pi = \frac{28}{10}\left[1 + \frac{2}{3}\left(\frac{2}{100}\right) + \frac{2\cdot 4}{3\cdot 5}\left(\frac{2}{100}\right)^2 + \cdots\right] + $$
$$+ \frac{30\,336}{100\,000}\left[1 + \frac{2}{3}\left(\frac{144}{100\,000}\right) + \frac{2\cdot 4}{3\cdot 5}\left(\frac{144}{100\,000}\right)^2 + \cdots\right] \qquad (16.16)$$

Euler, 1779 [114, p. 295]

$$\pi = 4\int_0^1 \sqrt{1-x^2}\,dx \qquad (16.17)$$

Euler, 1738

$$\pi = \frac{2\cdot 6}{5}\int_0^1 \frac{dx}{\sqrt{1-x^4}} \qquad (16.18)$$

Euler, 1739

$$\pi = 4\int_0^1 \frac{dx}{\sqrt{1-x^4}}\int_0^1 \frac{x^2\,dx}{\sqrt{1-x^4}} \qquad (16.19)$$

Euler, 1748

$$\pi = \lim_{n\to\infty} \frac{4}{n^2}\sum_{k=0}^n \sqrt{n^2-k^2} \qquad (16.20)$$

Euler, 1738

$$\pi = \lim_{n\to\infty}\left[\frac{1}{n} + \frac{1}{6n^2} + 4n\left(\frac{1}{n^2+1^2} + \frac{1}{n^2+2^2} + \cdots + \frac{1}{n^2+n^2}\right)\right] \qquad (16.21)$$

Euler [41, (3-12)]

$$\pi = 3\sqrt{3}\left(1 - \frac{1}{2} + \frac{1}{4} - \frac{1}{5} + \frac{1}{7} - \frac{1}{8} + \cdots + \frac{1}{3n+1} - \frac{1}{3n+2} + \cdots\right) \qquad (16.22)$$

Euler, 1739

$$\frac{1}{\pi} = \sum_{n=1}^\infty \frac{1}{2^{n+1}}\tan\frac{\pi}{2^{n+1}} \qquad (16.23)$$

Euler

16. π Formula Collection

$$\frac{1}{\pi^2} = \frac{1}{6}\left(1 - \frac{1}{2^2} - \frac{1}{3^2} - \frac{1}{5^2} + \frac{1}{(2\cdot 3)^2} - \frac{1}{7^2} + \frac{1}{(2\cdot 5)^2} - \frac{1}{11^2} - \frac{1}{13^2} + \right. \quad (16.24)$$

$$+ \frac{1}{(2\cdot 7)^2} + \frac{1}{(3\cdot 5)^2} - \frac{1}{17^2} - \frac{1}{19^2} + \frac{1}{(3\cdot 7)^2} + \frac{1}{(2\cdot 11)^2} - \quad (16.25)$$

$$\left. - \frac{1}{23^2} + \frac{1}{(2\cdot 13)^2} - \frac{1}{29^2} - \frac{1}{(2\cdot 3\cdot 5)^2} - \cdots\right) \quad (16.26)$$

Euler, 1737

$$\frac{\pi^2}{6} = \frac{1}{1^2} + \frac{1}{2^2} + \frac{1}{3^2} + \frac{1}{4^2} + \frac{1}{5^2}\cdots \quad (16.27)$$

Euler, 1736

$$\frac{\pi^2}{6} = \frac{2^2}{2^2-1}\cdot\frac{3^2}{3^2-1}\cdot\frac{5^2}{5^2-1}\cdot\frac{7^2}{7^2-1}\cdots \quad (16.28)$$

$$\frac{\pi^2}{8} = \frac{1}{1^2} + \frac{1}{3^2} + \frac{1}{5^2} + \frac{1}{7^2}\cdots \quad (16.29)$$

Euler, 1748

$$\frac{\pi^2}{12} = 1 - \frac{1}{2^2} + \frac{1}{3^2} - \frac{1}{4^2} + \frac{1}{5^2}\cdots \quad (16.30)$$

Euler, 1748

$$\frac{\pi^3}{32} = \frac{1}{1^3} - \frac{1}{3^3} + \frac{1}{5^3} - \frac{1}{7^3} \pm \cdots \quad (16.31)$$

Euler, 1748

$$\pi^4 = \frac{9}{680}\sum_{k=1}^{\infty}\frac{1}{k^4\binom{2k}{k}} \quad (16.32)$$

Euler, 1738

$$\frac{\pi^4}{90} = \frac{1}{1^4} + \frac{1}{2^4} + \frac{1}{3^4} + \cdots \quad (16.33)$$

Euler, 1748

$$\frac{\pi^4}{96} = 1 + \frac{1}{3^4} + \frac{1}{5^4} + \frac{1}{7^4} + \cdots \quad (16.34)$$

Euler, 1748

$$\frac{7\pi^4}{720} = 1 - \frac{1}{2^4} + \frac{1}{3^4} - \frac{1}{4^4} + \cdots \quad (16.35)$$

Euler, 1748

$$\pi^{26} = \frac{1}{76977927}\frac{27!}{2^{24}}\left(\frac{1}{1^{26}} + \frac{1}{2^{26}} + \frac{1}{3^{26}} + \cdots\right) \quad (16.36)$$

Euler, 1738

$$\pi = \frac{1}{n} + 4n\left(\frac{1}{n^2+1^2} + \frac{1}{n^2+2^2} + \cdots + \frac{1}{n^2+n^2}\right) - \frac{4\pi}{e^{2\pi n}} +$$

$$+ \frac{B_2}{n^2} - \frac{B_6}{3\cdot 2^2\cdot n^6} + \frac{B_{10}}{5\cdot 2^4\cdot n^{10}} - \frac{B_{14}}{7\cdot 2^6\cdot n^{14}} + \cdots \quad (16.37)$$

where the B_i are the Bernoulli numbers: $B_2 = 1/6$, $B_6 = 1/42\ldots$

Euler [41, (3-13)]

$$\frac{6}{\pi^2} = \prod_{\substack{p=2 \\ p\ prime}}^{\infty}\left(1 - \frac{1}{p^2}\right) \quad (16.38)$$

Euler, 1748

$$\frac{\pi}{2} = \frac{3}{2} \cdot \frac{5}{6} \cdot \frac{7}{6} \cdot \frac{11}{10} \cdot \frac{13}{14} \cdot \frac{17}{18} \cdot \frac{19}{18} \cdots \qquad (16.39)$$

where the numerators are the odd prime numbers
and the denominators are even numbers, *not* divisible by 4,
which differ from the numerators by 1.

Euler [20, p. 656]

$$\frac{\pi}{2} = 1 + \frac{1}{1} + \frac{2}{1} + \frac{6}{1} + \frac{12}{1} + \frac{20}{1} + \frac{30}{1} + \frac{42}{1} + \cdots \qquad (16.40)$$

Euler, 1739

$$\frac{3}{4}\pi = 2 + \frac{1}{2} + \frac{3}{2} + \frac{8}{2} + \frac{15}{2} + \frac{24}{2} + \cdots \qquad (16.41)$$

Euler, 1739

$$\frac{\sin \pi z}{\pi} = z \prod_{k=1}^{\infty} \left(1 - \frac{z^2}{k^2}\right) \qquad (16.42)$$

For $z = 1/2$, (16.91) applies.

Euler, 1735

$$\frac{\pi}{\sin \pi z} = \Gamma(z)\Gamma(1-z) \qquad z \text{ real, non-integer} \qquad (16.43)$$

where $\Gamma(x) := \int_0^\infty e^{-t} t^{x-1} dt$ for all real $x > 0$
Euler's Gamma function

Formulae of S. Ramanujan (1887–1920)

$$\frac{1}{\pi} = \sum_{n=0}^{\infty} \binom{2n}{n}^3 \frac{42n+5}{2^{12n+4}} \qquad (16.44)$$

Ramanujan [96, (29)], 1914

$$\frac{1}{\pi} = \frac{1}{3528} \sum_{n=0}^{\infty} (-1)^k \frac{(4n)!!}{(n!)^4 4^{4k}} \frac{[1123 + 21460n]}{882^{2n}} \qquad (16.45)$$

Ramanujan [96, (39)], 1914

$$\frac{1}{\pi} = \frac{\sqrt{8}}{9801} \sum_{n=0}^{\infty} \frac{(4n)!}{(n!)^4} \frac{[1103 + 26390n]}{396^{4n}} \qquad (16.46)$$

Ramanujan [96, (44)], 1914

$$\frac{\pi}{2} = 1 + \frac{1}{2}\left(\frac{1}{3}\right) + \frac{1 \cdot 3}{2 \cdot 4}\left(\frac{1}{5}\right) + \frac{1 \cdot 3 \cdot 5}{2 \cdot 4 \cdot 6}\left(\frac{1}{7}\right) + \cdots \qquad (16.47)$$

Ramanujan [41, (8-1)]

$$\sqrt{\frac{1}{2}\pi e} = 1 + \frac{1}{1 \cdot 3} + \frac{1}{1 \cdot 3 \cdot 5} + \frac{1}{1 \cdot 3 \cdot 5 \cdot 7} + \cdots + \\ + \frac{1}{1} + \frac{1}{1} + \frac{2}{1} + \frac{3}{1} + \frac{4}{1} + \cdots \qquad (16.48)$$

Ramanujan [41, p. 89]

$$\frac{\pi}{2}\ln 2 = 1 + \frac{1}{2}\cdot\frac{1}{3^2} + \frac{1\cdot 3}{2\cdot 4}\cdot\frac{1}{5^2} + \frac{1\cdot 3\cdot 5}{2\cdot 4\cdot 6}\cdot\frac{1}{7^2} + \cdots \qquad (16.49)$$
Ramanujan [41, (8-2)]

$$\frac{\pi^3}{48} + \frac{\pi}{4}(\ln 2)^2 = 1 + \frac{1}{2}\cdot\frac{1}{3^3} + \frac{1\cdot 3}{2\cdot 4}\cdot\frac{1}{5^3} + \frac{1\cdot 3\cdot 5}{2\cdot 4\cdot 6}\cdot\frac{1}{7^3} + \cdots \qquad (16.50)$$
Ramanujan [41, (8-3)]

BBP-like series

$$\pi = \sum_{n=0}^{\infty} \frac{(-1)^n}{4^n}\left(\frac{2}{4n+1} + \frac{2}{4n+2} + \frac{1}{4n+3}\right) \qquad (16.51)$$
Adamchik and Wagon, 1997 [2]

$$\pi = \sum_{n=0}^{\infty} \frac{1}{16^n}\left(\frac{4}{8n+1} - \frac{2}{8n+4} - \frac{1}{8n+5} - \frac{1}{8n+6}\right) \qquad (16.52)$$
Borwein, Bailey and Plouffe, 1995

$$\pi = \frac{1}{2^6}\sum_{n=0}^{\infty}\frac{(-1)^n}{2^{10n}}\left(-\frac{2^5}{4n+1} - \frac{1}{4n+3} + \frac{2^8}{10n+1} + \right.$$
$$\left. - \frac{2^6}{10n+3} - \frac{2^2}{10n+5} - \frac{2^2}{10n+7} + \frac{1}{10n+9}\right) \qquad (16.53)$$
Bellard, 1996 [19]

$$\pi^2 = \sum_{n=1}^{\infty}\frac{1}{n^3}\left(\frac{-12}{n+1} + \frac{384}{n+2} + \frac{45/2}{2n+1} - \frac{1215/2}{2n+3}\right) \qquad (16.54)$$
Adamchik and Wagon, 1997 [2]

$$\pi = \sum_{n=1}^{\infty}\frac{1}{n^3}\left(\frac{-238}{n+1} + \frac{285/2}{2n+1} - \frac{667/32}{4n+1} - \frac{5103/16}{4n+3} + \right.$$
$$\left. + \frac{35625/32}{4n+5}\right) \qquad (16.55)$$
Adamchik and Wagon, 1997 [2]

$$\pi = \frac{1}{2^{26}}\sum_{n=0}^{\infty}\frac{(-1)^n}{2^{30n}}\left(\frac{2^{27}}{20n+1} + \frac{2^{25}}{12n+1} + \frac{2^{25}}{10n+1} + \frac{2^{24}}{20n+3} + \right.$$
$$+ \frac{2^{22}}{6n+1} - \frac{2^{20}}{60n+15} - \frac{2^{19}}{10n+3} - \frac{2^{18}}{20n+7} - \frac{2^{15}}{12n+5} +$$
$$+ \frac{2^{15}}{20n+9} + \frac{2^{12}}{30n+15} + \frac{2^{12}}{20n+11} - \frac{2^{10}}{12n+7} - \frac{2^9}{20n+13} -$$
$$- \frac{2^7}{10n+7} - \frac{2^5}{60n+45} + \frac{2^2}{6n+5} + \frac{2^3}{20n+17} + \frac{2}{10n+9} +$$
$$\left. + \frac{1}{12n+11} + \frac{1}{12n+19}\right) \qquad (16.56)$$
Pschill, 1999

$$\pi = \frac{2}{\sqrt{3}} \sum_{n=0}^{\infty} \frac{1}{9^n} \left(\frac{3}{4n+1} - \frac{1}{4n+3} \right) \tag{16.57}$$

or, general:

$$\pi = \frac{2}{3^{j-2}\sqrt{3}} \sum_{n=0}^{\infty} \frac{(-1)^{jn}}{3^{jn}} \left(\sum_{l=1}^{j} \frac{(-1)^{l-1} 3^{j-l}}{2jn + 2l - 1} \right), j \geq 2, 3, 4 \ldots \tag{16.58}$$

Pschill, 1999

Other series

$$\pi = 4 \left(1 - \frac{1}{3} + \frac{1}{5} - \frac{1}{7} + - \cdots \right) \tag{16.59}$$

Leibniz, 1674

$$\pi = 2\sqrt{2} \left(1 + \frac{1}{3} - \frac{1}{5} - \frac{1}{7} + \frac{1}{9} + \frac{1}{11} - - + + \cdots \right) \tag{16.60}$$

$$\pi = \lim_{n \to \infty} \left(\frac{2^{2n}}{\binom{2n}{n}} \right)^2 \cdot \frac{1}{n} \tag{16.61}$$

Knopp after [41]

$$\pi = \lim_{n \to \infty} \left(\frac{2^{2n}}{\binom{2n}{n}} \right)^2 \cdot \frac{1}{(n + \frac{1}{4})} \tag{16.62}$$

Bauer [17], converges a lot better

$$\pi = 24 \left(\frac{\sqrt{3}}{32} + \int_0^{1/4} \sqrt{x - x^2} \, dx \right) \tag{16.63}$$

$$= \frac{3\sqrt{3}}{4} + 24 \left(\frac{1}{12} - \frac{1}{5 \cdot 2^5} - \frac{1}{28 \cdot 2^7} - \frac{1}{72 \cdot 2^9} - \right.$$

$$\left. - \sum_{k=4}^{\infty} \frac{3 \cdot 5 \cdots (2k-3)}{4 \cdot 6 \cdots (2k) \cdot (2k+3)} \frac{1}{2^{2k+3}} \right) \tag{16.64}$$

Newton, 1665, after [32, p. 339]

$$\frac{\pi}{6} = \arcsin \frac{1}{2} \tag{16.65}$$

$$= \frac{1}{2} + \frac{1}{2} \cdot \frac{1}{3 \cdot 2^3} + \frac{1 \cdot 3}{2 \cdot 4} \cdot \frac{1}{5 \cdot 2^5} + \frac{1 \cdot 3 \cdot 5}{2 \cdot 4 \cdot 6} \cdot \frac{1}{7 \cdot 2^7} + \cdots \tag{16.66}$$

Newton, 1676

$$\frac{\pi}{6} = \arctan \frac{\sqrt{3}}{3} \tag{16.67}$$

$$= \frac{\sqrt{3}}{3} \left(1 - \frac{1}{3 \cdot 3} + \frac{1}{3^2 \cdot 5} - \frac{1}{3^3 \cdot 7} + \frac{1}{3^4 \cdot 9} - \cdots \right) \tag{16.68}$$

Sharp, 1699 [114, p. 294]

$$\pi = 16\sqrt{3} \left(\frac{1}{1 \cdot 3 \cdot 3} + \frac{2}{5 \cdot 7 \cdot 3^3} + \frac{3}{9 \cdot 11 \cdot 3^5} + \cdots \right) \tag{16.69}$$

Used by De Lagny in 1719

$$\frac{\pi}{2} = 1 + \sum_{s=0}^{\infty} \frac{2 \cdot 4 \cdot 6 \cdots 2s}{3 \cdot 5 \cdot 7 \cdots (2s+1)} \cdot \frac{1}{2^s} \tag{16.70}$$

After Stirling, 1730, [85, p. 81]

$$\frac{\pi^2}{12} = \frac{(\ln 2)^2}{2} + \sum_{s=1}^{\infty} \frac{1}{s^2} \cdot \frac{1}{2^s} \tag{16.71}$$

Legendre, 1811, [85, p. 81]

$$\pi = \sum_{x=-\infty}^{+\infty} \left(\frac{\sin x}{x} \right) \tag{16.72}$$

$$\pi = \sum_{x=-\infty}^{+\infty} \left(\frac{\sin x}{x} \right)^2 \tag{16.73}$$

$$\pi = \frac{3}{4} \sum_{x=-\infty}^{+\infty} \left(\frac{\sin x}{x} \right)^3 \tag{16.74}$$

$$\pi = \frac{3}{2} \sum_{x=-\infty}^{+\infty} \left(\frac{\sin x}{x} \right)^4 \tag{16.75}$$

$$\frac{4}{\pi} = 1 + \frac{1}{2}\left(\frac{1}{2}\right)^2 + \frac{1}{3}\left(\frac{1 \cdot 3}{2 \cdot 4}\right)^2 + \frac{1}{4}\left(\frac{1 \cdot 3 \cdot 5}{2 \cdot 4 \cdot 6}\right)^2 + \cdots \tag{16.76}$$

Catalan, 1948 [41, p. 86]

$$\pi = 3 + \frac{1}{6}\left(\sum_{n=1}^{\infty} \frac{(-1)^{n+1}}{\sum_{k=1}^{n} k^2}\right) \tag{16.77}$$

J. and P. Borwein [32, p. 101]

$$\pi = \frac{1}{740025}\left(\sum_{n=1}^{\infty} \frac{3 P(n)}{\binom{7n}{2n} 2^{n-1}}\right) - 20379280 \tag{16.78}$$

where $P(n) := -885673181\, n^5 + 3125347237\, n^4 -$
$- 2942969225\, n^3 + 1031962795\, n^2 -$
$- 196882274\, n + 10996648$

Bellard, 1996 [19]

$$\pi = \sum_{n=0}^{\infty} \frac{n!^2 2^{n+1}}{(2n+1)!} \tag{16.79}$$

$$= 2 + \frac{1}{3}(2 + \frac{2}{5}(2 + \frac{3}{7}(2 + \frac{4}{9}(2 + \cdots)))) \tag{16.80}$$

Series for spigot algorithm (see page 78)

$$\pi = 3 + \frac{1}{60}(8 + \frac{2 \cdot 3}{7 \cdot 8 \cdot 3}(13 + \frac{3 \cdot 5}{10 \cdot 11 \cdot 3} \cdot (18 + \frac{4 \cdot 7}{13 \cdot 14 \cdot 3}(23 + \cdots)))) \tag{16.81}$$

Gosper

$$\pi = \sum_{n=0}^{\infty} \frac{(25n-3)n!(2n)!}{2^{n-1}(3n)!} \tag{16.82}$$

Gosper, 1974

$$\pi = \sqrt{\frac{297}{25} \sum_{n=1}^{\infty} b(n) \left(\frac{1}{n^2} - \frac{1}{(n+1)^2}\right)} \qquad (16.83)$$

where $b(n)$ refers to the number of odd decimal digits in n
($b(901) = 2$, $b(811) = 2$, $b(406) = 0$)

J. and P. Borwein, 1992[34]

$$\pi = 3\sqrt{3} \sum_{n=1}^{\infty} \frac{1}{n\binom{2n}{n}} \qquad (16.84)$$

Comtet, 1974

$$\pi^2 = 18 \sum_{n=1}^{\infty} \frac{1}{n^2 \binom{2n}{n}} \qquad (16.85)$$

Euler, 1748

$$\pi^2 = 4\left[1 + \sum_{n=1}^{\infty} \frac{2^{2n+1}(n!)^2}{(2n+2)!}\right] \qquad (16.86)$$

Takebe Kenko (1722) [98, p. 306]

$$\pi^2 = 72 \sum_{n=1}^{\infty} \frac{(2-\sqrt{3})^n}{n^2 \binom{2n}{n}} \qquad (16.87)$$

Crandall, 1994 [47]

$$\pi^4 = \frac{3240}{17} \sum_{n=1}^{\infty} \frac{1}{n^4 \binom{2n}{n}} \qquad (16.88)$$

Comtet, 1974

$$\frac{1}{\pi} = \frac{12}{\sqrt{640320^3}} \sum_{k=0}^{\infty} (-1)^k \frac{(6k)!}{(k!)^3 (3k)!} \cdot \frac{13591409 + 545140134k}{(640320^3)^k} \qquad (16.89)$$

D. and G. Chudnovsky, 1987

Products

$$\frac{2}{\pi} = \sqrt{\frac{1}{2}} \cdot \sqrt{\frac{1}{2} + \frac{1}{2}\sqrt{\frac{1}{2}}} \cdot \sqrt{\frac{1}{2} + \frac{1}{2}\sqrt{\frac{1}{2} + \frac{1}{2}\sqrt{\frac{1}{2}}}} \cdots \qquad (16.90)$$

Viète, 1593

$$\frac{4}{\pi} = \frac{3 \cdot 3}{2 \cdot 4} \cdot \frac{5 \cdot 5}{4 \cdot 6} \cdot \frac{7 \cdot 7}{6 \cdot 8} \cdots = \prod_{n=1}^{\infty}\left(1 + \frac{1}{4n(n+1)}\right) \qquad (16.91)$$

$$= 2 \cdot \frac{1 \cdot 3}{2 \cdot 2} \cdot \frac{3 \cdot 5}{4 \cdot 4} \cdot \frac{5 \cdot 7}{6 \cdot 6} \cdot \frac{7 \cdot 9}{8 \cdot 8} \cdots = 2 \prod_{n=1}^{\infty}\left(1 - \frac{1}{4n^2}\right) \qquad (16.92)$$

Wallis, 1655

$$\frac{2}{\pi} = \prod_{n=1}^{p} \sqrt{\frac{1}{2} + \frac{1}{2}\sqrt{\frac{1}{2} + \frac{1}{2}\sqrt{\frac{1}{2} + \cdots + \frac{1}{2}\sqrt{\frac{1}{2}}}}} \times \prod_{n=1}^{\infty} \frac{2^{p+1}n - 1}{2^{p+1}n} \cdot \frac{2^{p+1}n + 1}{2^{p+1}n}$$

(n radicals)
$$\qquad (16.93)$$

Osler[88], 1999

$$e^\pi = 32 \prod_{j=0}^{\infty} \left(\frac{a_{j+1}}{a_j}\right)^{2^{-j+1}} \tag{16.94}$$

where $a_0 = 1$, $b_0 = 1/\sqrt{2}$, $a_{j+1} = (a_j + b_j)/2$, $b_{j+1} = \sqrt{a_j \cdot b_j}$

Salamin [100, p. 569]

Continued fractions

For an explanation of the notation, see (4.67) on page 65.

$$\pi = 3 + \frac{1}{7+} \frac{1}{15+} \frac{1}{1+} \frac{1}{292+} \frac{1}{1+} \frac{1}{1+} \frac{1}{1+} \frac{1}{2+} \frac{1}{1+} \frac{1}{3+} \frac{1}{1+} \frac{1}{14+} \cdots \tag{16.95}$$

$$\frac{4}{\pi} = 1 + \frac{1^2}{2+} \frac{3^2}{2+} \frac{5^2}{2+} \frac{7^2}{2+} \frac{9^2}{2+} \cdots \tag{16.96}$$

Brouncker, 1658

$$\frac{4}{\pi} = 1 + \frac{2}{7+} \frac{1 \cdot 3}{8+} \frac{3 \cdot 5}{8+} \frac{5 \cdot 7}{8+} \cdots \tag{16.97}$$

Euler, 1783 [114, p. 292]

$$\frac{\pi}{2} = 1 + \frac{2}{3+} \frac{1 \cdot 3}{4+} \frac{3 \cdot 5}{4+} \frac{5 \cdot 7}{4+} \cdots \tag{16.98}$$

Euler, 1783 [114, p. 292]

$$\frac{4}{\pi} = 1 + \frac{1^2}{3+} \frac{2^2}{5+} \frac{3^2}{7+} \frac{4^2}{9+} \cdots \tag{16.99}$$

$$\frac{\pi^2}{6} = 1 + \frac{1}{1+} \frac{1 \cdot 1}{1+} \frac{1 \cdot 2}{1+} \frac{2 \cdot 2}{1+} \frac{2 \cdot 3}{1+} \frac{3 \cdot 3}{1+} \frac{3 \cdot 4}{1+} \cdots \tag{16.100}$$

$$\frac{12}{\pi^2} = 1 + \frac{1^4}{3+} \frac{2^4}{5+} \frac{3^4}{7+} \frac{5^4}{9+} \cdots \tag{16.101}$$

$$\frac{\pi}{2} = 1 - \frac{1}{3-} \frac{2 \cdot 3}{1-} \frac{1 \cdot 2}{3-} \frac{4 \cdot 5}{1-} \cdots \tag{16.102}$$

Stern, 1833

$$\pi = 3 + \frac{1^2}{6+} \frac{3^2}{6+} \frac{5^2}{6+} \frac{7^2}{6+} \cdots \tag{16.103}$$

Lange, 1999 [80]

Arctan relations

With the notation $[x] = \arctan \frac{1}{x}$:

$$\frac{\pi}{4} = [1] \tag{16.104}$$

$$\frac{\pi}{2} = 2[\sqrt{2}] + [2\sqrt{2}] \tag{16.105}$$
Wetherfield, 1996

$$\frac{\pi}{6} = 2[3\sqrt{3}] + [4\sqrt{3}] \tag{16.106}$$

$$\frac{\pi}{4} = [2] + [3] \tag{16.107}$$
Euler, 1738

$$\frac{\pi}{4} = [2] + [5] + [8] \tag{16.108}$$
Strassnitzky, 1844

$$\frac{\pi}{4} = 2[2] - [7] \tag{16.109}$$
Hermann, 1706

$$\frac{\pi}{4} = 2[3] + [7] \tag{16.110}$$
Hutton, 1776

$$\frac{\pi}{4} = 3[4] + [20] + [1985] \tag{16.111}$$
Loney, 1893

$$\frac{\pi}{4} = 4[5] - [70] + [99] \tag{16.112}$$
Euler, 1764

$$\frac{\pi}{4} = 4[5] - [239] \tag{16.113}$$
Machin, 1706

$$\frac{\pi}{4} = 5[7] + 2[79/3] \tag{16.114}$$
Euler, 1755 [114, p. 295]

$$\frac{\pi}{4} = 6[8] + 2[57] + [239] \tag{16.115}$$
Størmer, 1896

$$\frac{\pi}{4} = 7[10] + 8[100] + [682] + 4[1000] + 3[1303] - 4[90109] - 2[500150] \tag{16.116}$$
Wrench Jr. [41, (9-34)]

$$\frac{\pi}{4} = 8[10] - [239] - 4[515] \tag{16.117}$$
Klingenstierna, c. 1730 [114, p. 296]

$$\frac{\pi}{4} = 8[101/10] - [239] + 4[52525] \tag{16.118}$$
Gauss [56, II, p. 524]

$$\frac{\pi}{4} = 12[18] + 8[57] - 5[239] \tag{16.119}$$
Gauss [56, II, p. 524]

$$\frac{\pi}{4} = 22[28] + 2[443] - 5[1393] - 10[11018] \tag{16.120}$$
Escott

$$\frac{\pi}{4} = 12[38] + 20[57] + 7[239] + 24[268] \tag{16.121}$$
Gauss [56, II, p. 525]

$$\frac{\pi}{4} = 44[57] + 7[239] - 12[682] + 24[12943] \qquad (16.122)$$
Størmer, 1896

$$\frac{\pi}{4} = 88[172] + 51[239] + 32[682] + 44[5357] + 68[12943] \qquad (16.123)$$
Størmer, 1896

$$\frac{\pi}{4} = 88[192] + 39[239] + 100[515] - 32[1068] - 56[173932] \qquad (16.124)$$
Arndt [7], 1993

$$\frac{\pi}{4} = 322[577] + 76[682] + 139[1393] + 156[12943] +$$
$$+ 132[32807] + 44[1049433] \qquad (16.125)$$
Arndt [7], 1993

$$\frac{\pi}{4} = 1587[2852] + 295[4193] + 593[4246] + 359[39307] +$$
$$+ 481[55603] + 625[211050] - 708[390112] \qquad (16.126)$$
Arndt [7], 1993

$$\frac{\pi}{4} = 1074[4246] + 1257[5357] + 1731[6107] + 295[12943] +$$
$$+ 625[19703] - 481[32807] - 1042[39307] + 398[390112] \qquad (16.127)$$
Arndt [7], 1993

$$\frac{\pi}{4} = 7162[12943] + 3796[32807] + 2558[34208] +$$
$$+ 2729[44179] - 708[51387] + 2192[114669] -$$
$$- 2805[157318] - 3696[485298] - 2407[24208144] \qquad (16.128)$$
Arndt [7], 1993

$$\frac{\pi}{4} = 50539[51387] + 1555[114669] - 6601[157318] -$$
$$- 20678[390112] - 5617[485298] - 64126[617427] +$$
$$+ 10958[1984933] - 30569[3449051] +$$
$$+ 23407[22709274] + 25433[24208144] \qquad (16.129)$$
Arndt [7], 1993

$$\frac{\pi}{4} = 36462[390112] + 135908[485298] + 274509[683982] -$$
$$- 39581[1984933] + 178477[2478328] - 114569[3449051] -$$
$$- 146571[18975991] + 61914[22709274] - 6044[24208144] -$$
$$- 89431[201229582] - 43938[2189376182] \qquad (16.130)$$
Arndt [7], 1993

$$\frac{\pi}{4} = 446879[683982] + 172370[1635786] - 193720[1984933] +$$
$$+ 369078[2478328] + 18231[3014557] + 21339[3449051] -$$
$$- 154139[6225244] - 110109[18975991] + 80145[22709274] -$$
$$- 223183[24208144] - 107662[201229582] - 216308[2189376182] \qquad (16.131)$$
Arndt [7], 1993

$$\frac{\pi}{4} = 872408[1984933] + 619249[2298668] + 369078[2478328] +$$
$$+ 18231[3014557] - 1217159[5033696] + 911989[6225244] +$$
$$+ 783649[18975991] - 70886[22709274] - 374214[24208144] -$$
$$- 1044789[168623905] + 339217[201229582] - 446879[2848622638] +$$
$$+ 402941[2189376182] \qquad (16.132)$$
Arndt, [7], 1993

$$\frac{\pi}{4} = \sum_{n=1}^{\infty} [n2 + n + 1] \qquad (16.133)$$
Knopp

$$\frac{\pi}{4} = \sum_{n=1}^{\infty} [F_{2n+1}] \tag{16.134}$$

where F_i: Fibonacci numbers, $F_1 = 1, F_2 = 1, F_{i+2} = F_{i+1} + F_i, i \geq 1$

Arndt, [7], 1994

$$\frac{\pi}{4} = \frac{3\sqrt{5} - 5}{2} - \sum_{n=1}^{\infty} F_{2n} \left[(3F_{2n+2} + F_{2n+2}^3)/2 \right] \tag{16.135}$$

Arndt, [7], 1994

Miscellaneous formulae

$$\pi = 2i \ln \frac{1-i}{1+i} \tag{16.136}$$

Fagnano [118, p. 199]

$$\pi = [2(1/2)!]^2 = (1/2)!(-1/2)!(-1/2) = \Gamma^2(1/2) = (1/2\Gamma(-1/2))^2 \tag{16.137}$$

where $\Gamma(x) := \int_0^\infty e^{-t} t^{x-1} dt$ for all real $x > 0$

Euler's Gamma function

$$\frac{\pi^2}{6} = \zeta(2) \tag{16.138}$$

where $\zeta(s) = \sum_{n=1}^{\infty} \frac{1}{n^s}$ Riemann's zeta function

$$\pi = \lim_{n \to \infty} \sqrt{\frac{6 \ln \prod_{i=1}^{n}(F_i)}{\ln lcm(F_1, \ldots, F_n)}} \tag{16.139}$$

where lcm: least common multiple,

F_i: Fibonacci numbers, $F_1 = 1, F_2 = 1, F_{i+2} = F_{i+1} + F_i, i \geq 1$

Matiyasevich [84]

$$\pi = \frac{22}{7} - \int_0^1 \frac{x^4(1-x)^4}{1+x^2} dx \tag{16.140}$$

Backhouse, [8], 1995

$$\frac{e+1}{e-1} = 2 + 4 \sum_{r=1}^{\infty} \frac{1}{(2\pi r)^2 + 1} \tag{16.141}$$

Kempermann 1995 [73, p. 98]

$$\begin{bmatrix} 0 & \pi + 6 \\ 0 & 1 \end{bmatrix} = \prod_{k=0}^{\infty} \begin{bmatrix} \frac{2(k-\frac{1}{2})(k+2)}{27(k+\frac{2}{3})(k+\frac{4}{3})} & 10 \\ 0 & 1 \end{bmatrix} \tag{16.142}$$

Bellard, 1996 [19]

$$\frac{17\pi^4}{360} = \sum_{k=1}^{\infty} \left(1 + \frac{1}{2} + \ldots + \frac{1}{k}\right)^2 \cdot k^{-2} \qquad (16.143)$$

Au-Yeung, 1993, after http://www.nas.nasa.gov/Pubs/NASnews/97/05/math.html

$$\frac{37\pi^6}{22680} = \zeta^2(3) + \sum_{k=1}^{\infty} \left(1 + \frac{1}{2} + \ldots + \frac{1}{k}\right)^2 \cdot k^{-4} \qquad (16.144)$$

Bailey, http://www.nas.nasa.gov/Pubs/NASnews/97/05/math.html

Iterative algorithms

Algorithm (16.145) Archimedes, 250 BC, linear (alg 16.145)

Initialise:
$$a_0 := \sqrt{3}$$
$$b_0 := 2$$

Iterate:
$$a_{k+1} := a_k + b_k \quad \xrightarrow{1} \quad 6 \cdot 2^{k+1}/\pi-$$
$$b_{k+1} := \sqrt{1 + a_{k+1}^2} \quad \xrightarrow{1} \quad 6 \cdot 2^{k+1}/\pi+$$

Then:
$$\frac{b_{k+1}}{6 \cdot 2^{k+1}} < \pi < \frac{a_{k+1}}{6 \cdot 2^{k+1}}$$

Archimedes, c. 250 BC

Algorithm (16.146) Descartes, c. 1649, linear (alg 16.146)

Initialise:
$$p_0 := \sqrt{2}$$
$$q_0 := 1$$

Iterate:
$$p_{k+1} := \frac{p_k + q_k}{2} \quad \xrightarrow{1} \quad \frac{4}{\pi}$$
$$q_{k+1} := \sqrt{p_{k+1} q_k} \quad \xrightarrow{1} \quad \frac{4}{\pi}$$

Then:
$$\frac{4}{q_{k+1}} < \pi < \frac{4}{p_{k+1}}$$

Descartes, c. 1649 [114, p. 289]

Algorithm (16.147) Bit recursion (alg 16.147)

Initialise:
$$a_0 := \tan(1)$$

Iterate:
$$a_{k+1} := (2 a_k)/(1 - a_k^2)$$
$$b(x) := \begin{cases} 1 & \text{If } x < 0 \\ 0 & \text{Else} \end{cases}$$

Then:
$$\frac{1}{\pi} = \sum_{k=0}^{\infty} \frac{b(a_k)}{2^{k+1}}$$

Plouffe from [36]

Algorithm (16.148) Gauss AGM (alg 16.148)

Initialise:
$$a_0 := 1$$
$$b_0 := 1/\sqrt{2}$$
$$s_0 := 0.5$$

Iterate:
$$a_{k+1} := (a_k + b_k)/2$$
$$b_{k+1} := \sqrt{a_k b_k}$$
$$c_{k+1}^2 := a_{k+1}^2 - b_{k+1}^2 = (a_{k+1} - a_k)^2$$
$$s_{k+1} := s_k - 2^{k+1} c_{k+1}^2$$

Then:
$$p_n = \frac{(a_n + b_n)^2}{2s_n} \xrightarrow{2} \pi$$
$$\pi - p_n = \frac{\pi 2^{n+4} e^{-\pi 2^{n+1}}}{\text{AGM}^2(1, 1/\sqrt{2})}$$

Gauss, c. 1800, Brent, 1976, Salamin, 1976

Algorithm (16.149) Gauss AGM, Schönhage variant (alg 16.149)

Initialise:
$$a_0 := 1$$
$$A_0 := 1$$
$$B_0 := 0.5$$
$$s_0 := 0.5$$

Iterate:
$$T := (A_k + B_k)/4$$
$$b_k := \sqrt{B_k}$$
$$a_{k+1} := (a_k + b_k)/2$$
$$A_{k+1} := a_{k+1}^2$$
$$B_{k+1} := 2(A_{k+1} - T)$$
$$s_{k+1} := s_k + 2^{k+1}(B_{k+1} - A_{k+1})$$

Then:
$$p_n = \frac{A_n + B_n}{s_n} \xrightarrow{2} \pi$$

Schönhage [105, p. 266]

Algorithm (16.150) Borwein, quadratic (alg 16.150)

Initialise:
$$a_0 := \sqrt{2}$$
$$b_0 := 0$$
$$p_0 := 2 + \sqrt{2}$$

Iterate:
$$a_{k+1} := (\sqrt{a_k} + \frac{1}{\sqrt{a_k}})/2$$
$$b_{k+1} := (\sqrt{a_k}(1 + b_k))/(a_k + b_k)$$
$$p_{k+1} := (p_k b_{k+1}(1 + a_{k+1}))/(1 + b_{k+1}) \xrightarrow{2} \pi$$

Then:
$$|p_n - \pi| \leq \frac{1}{10^{2^n}}$$

J. and P. Borwein, 1984 [30, p. 360]

Algorithm (16.151) Borwein, cubic (alg 16.151)

Initialise:
$$a_0 := 1/3$$
$$s_0 := (\sqrt{3} - 1)/2$$

Iterate:
$$r_{k+1} := 3/(1 + 2(1 - s_k^3)^{1/3})$$
$$s_{k+1} := (r_{k+1} - 1)/2$$
$$a_{k+1} := r_{k+1}^2 - 3^k(r_{k+1}^2 - 1) \xrightarrow{3} \frac{1}{\pi}$$

J. and P. Borwein, 1987 [13, p. 53]

Algorithm (16.152) Borwein, quartic (alg 16.152)

Initialise:
$$a_0 := 6 - 4\sqrt{2}$$
$$y_0 := \sqrt{2} - 1$$

Iterate:
$$y_{k+1} := ((1 - y_k^4)^{-1/4} - 1)/((1 - y_k^4)^{-1/4} + 1) \to 0+$$
$$a_{k+1} := a_k((1 + y_{k+1})^2)^2 - 2^{2k+3} y_{k+1}((1 + y_{k+1})^2 - y_{k+1}) \xrightarrow{4} \frac{1}{\pi}$$

Then:
$$0 < a_n - \pi^{-1} \leq 16 \cdot 4^k \cdot 2e^{-4^n 2\pi}$$

J. and P. Borwein, 1987 [32, p. 170]

Algorithm (16.153) Borwein, quintic (alg 16.153)

Initialise:
$$s_0 := 5(\sqrt{5} - 2)$$
$$a_0 := 1/2$$

Iterate:
$$s_{k+1} := (25)/(s_k(z + x/z + 1)^2) \xrightarrow{5} 1$$
where
$$x := \frac{5}{s_k} - 1$$
$$y := (x - 1)^2 + 7$$
$$z := \left(\frac{x}{2}\left(y + \sqrt{y^2 - 4x^3}\right)\right)^{1/5}$$
$$a_{k+1} = s_k^2 a_k - 5^k\left(\frac{s_k^2 - 5}{2} + \sqrt{s_k(s_k^2 - 2s_k + 5)}\right) \xrightarrow{5} \frac{1}{\pi}$$

Then:
$$a_n - \frac{1}{\pi} < 16 \cdot 5^n e^{-\pi 5^n}$$

J. and P. Borwein, 1987 [33, p. 115]

Algorithm (16.154) Borwein, nonic (alg 16.154)

Initialise:
$$a_0 := 1/3$$
$$r_0 := (\sqrt{3} - 1)/2$$
$$s_0 := (1 - r_0^3)^{1/3}$$

Iterate:
$$t := 1 + 2r_k$$
$$u := \left(9r_k(1 + r_k + r_k^2)\right)^{1/3}$$
$$v := t^2 + tu + u^2$$
$$m := (27(1 + s_k + s_k^2))/v$$
$$a_{k+1} := ma_k + 3^{2k-1}(1 - m) \xrightarrow{9} \frac{1}{\pi}$$
$$s_{k+1} := (1 - r_k)^3/((t + 2u)v)$$
$$r_{k+1} := (1 - s_k^3)^{1/3}$$

J. and P. Borwein, 1987 [13, p. 53]

17. Tables

17.1 Selected constants to 100 places (base 10)

π = 3.1415926535 8979323846 2643383279 5028841971 6939937510
       5820974944 5923078164 0628620899 8628034825 3421170679

$1/\pi$ = 0.3183098861 8379067153 7767526745 0287240689 1929148091
         2897495334 6881177935 9526845307 0180227605 5325061719

$6/\pi^2$ = 0.6079271018 5402662866 3276779258 3658334261 5264803347
           9293073654 1913650387 2577341264 7147255643 5537310256

π^2 = 9.8696044010 8935861883 4490999876 1511353136 9940724079
         0626413349 3762200448 2241920524 3001773403 7185522318

π^3 = 31.0062766802 9982017547 6315067101 3952022252 8856588510
          7694144538 1038063949 1746570603 7566701032 6028861930

π^4 = 97.4090910340 0243723644 0332688705 1112497275 8567268542
          1691467859 3899708554 5682719619 0121867234 7529925509

$\sqrt[2]{\pi}$ = 1.7724538509 0551602729 8167483341 1451827975 4945612238
                 7128213807 7898529112 8459103218 1374950656 7385446654

$\sqrt[3]{\pi}$ = 1.4645918875 6152326302 0142527263 7903917385 9685562793
                 7174357255 9371383936 4979828626 6145682067 8203538208

$\sqrt[4]{\pi}$ = 1.3313353638 0038971279 7534917950 2808533093 6622381810
                 4258453707 4828667007 6101723561 4968245891 0567069459

e = 2.7182818284 5904523536 0287471352 6624977572 4709369995
      9574966967 6277240766 3035354759 4571382178 5251664274

π^e = 22.4591577183 6104547342 7152204543 7350275893 1513399669
          2249203002 5540669260 4039911791 2318519752 7271430315

e^π = 23.1406926327 7926900572 9086367948 5473802661 0624260021
          1993445046 4095243423 5069045278 3516971997 0675492196

$e^{\pi/4}$ = 2.1932800507 3801545655 9769659278 7382234616 3764199427
              2334858015 9186570268 6418923693 4126522812 5781694047

$\ln \pi$ = 1.1447298858 4940017414 3427351353 0587116472 9481291531
           1571513623 0714721377 6988482607 9783623270 2754897077

$\log_2 \pi$ = 1.6514961294 7231879804 3279295108 0073350184 7692676304
              1529406788 5154881029 6358454143 8960264792 8098541017

$\log_{10} \pi$ = 0.4971498726 9413385435 1268288290 8988736516 7832438044
                 2446134053 4999249471 1208955267 4655547386 4642912223

$e^{\pi\sqrt{163}}$ = 2 6253741264 0768743.
                     9999999999 9925007259 7198185688 8793538563 3733699086
                     2707537410 3782106479 1011860731 2951181346 1860645041

17.2 Digits 0 to 2,500 of π (base 10)

```
π  =   3.1415926535 8979323846 2643383279 5028841971 6939937510
(0051)    5820974944 5923078164 0628620899 8628034825 3421170679
(0101)    8214808651 3282306647 0938446095 5058223172 5359408128
(0151)    4811174502 8410270193 8521105559 6446229489 5493038196
(0201)    4428810975 6659334461 2847564823 3786783165 2712019091
(0251)    4564856692 3460348610 4543266482 1339360726 0249141273
(0301)    7245870066 0631558817 4881520920 9628292540 9171536436
(0351)    7892590360 0113305305 4882046652 1384146951 9415116094
(0401)    3305727036 5759591953 0921861173 8193261179 3105118548
(0451)    0744623799 6274956735 1885752724 8912279381 8301194912

(0501)    9833673362 4406566430 8602139494 6395224737 1907021798
(0551)    6094370277 0539217176 2931767523 8467481846 7669405132
(0601)    0005681271 4526356082 7785771342 7577896091 7363717872
(0651)    1468440901 2249534301 4654958537 1050792279 6892589235
(0701)    4201995611 2129021960 8640344181 5981362977 4771309960
(0751)    5187072113 4999999837 2978049951 0597317328 1609631859
(0801)    5024459455 3469083026 4252230825 3344685035 2619311881
(0851)    7101000313 7838752886 5875332083 8142061717 7669147303
(0901)    5982534904 2875546873 1159562863 8823537875 9375195778
(0951)    1857780532 1712268066 1300192787 6611195909 2164201989

(1001)    3809525720 1065485863 2788659361 5338182796 8230301952
(1051)    0353018529 6899577362 2599413891 2497217752 8347913151
(1101)    5574857242 4541506959 5082953311 6861727855 8890750983
(1151)    8175463746 4939319255 0604009277 0167113900 9848824012
(1201)    8583616035 6370766010 4710181942 9555961989 4676783744
(1251)    9448255379 7747268471 0404753464 6208046684 2590694912
(1301)    9331367702 8989152104 7521620569 6602405803 8150193511
(1351)    2533824300 3558764024 7496473263 9141992726 0426992279
(1401)    6782354781 6360093417 2164121992 4586315030 2861829745
(1451)    5570674983 8505494588 5869269956 9092721079 7509302955

(1501)    3211653449 8720275596 0236480665 4991198818 3479775356
(1551)    6369807426 5425278625 5181841757 4672890977 7727938000
(1601)    8164706001 6145249192 1732172147 7235014144 1973568548
(1651)    1613611573 5255213347 5741849468 4385233239 0739414333
(1701)    4547762416 8625189835 6948556209 9219222184 2725502542
(1751)    5688767179 0494601653 4668049886 2723279178 6085784383
(1801)    8279679766 8145410095 3883786360 9506800642 2512520511
(1851)    7392984896 0841284886 2694560424 1965285022 2106611863
(1901)    0674427862 2039194945 0471237137 8696095636 4371917287
(1951)    4677646575 7396241389 0865832645 9958133904 7802759009

(2001)    9465764078 9512694683 9835259570 9825822620 5224894077
(2051)    2671947826 8482601476 9909026401 3639443745 5305068203
(2101)    4962524517 4939965143 1429809190 6592509372 2169646151
(2151)    5709858387 4105978859 5977297549 8930161753 9284681382
(2201)    6868386894 2774155991 8559252459 5395943104 9972524680
(2251)    8459872736 4469584865 3836736222 6260991246 0805124388
(2301)    4390451244 1365497627 8079771569 1435997700 1296160894
(2351)    4169486855 5848406353 4220722258 2848864815 8456028506
(2401)    0168427394 5226746767 8895252138 5225499546 6672782398
(2451)    6456596116 3548862305 7745649803 5593634568 1743241125
```

17.3 Digits 2,501 to 5,000 of π (base 10)

```
(2501)  1507606947  9451096596  0940252288  7971089314  5669136867
(2551)  2287489405  6010150330  8617928680  9208747609  1782493858
(2601)  9009714909  6759852613  6554978189  3129784821  6829989487
(2651)  2265880485  7564014270  4775551323  7964145152  3746234364
(2701)  5428584447  9526586782  1051141354  7357395231  1342716610
(2751)  2135969536  2314429524  8493718711  0145765403  5902799344
(2801)  0374200731  0578539062  1983874478  0847848968  3321445713
(2851)  8687519435  0643021845  3191048481  0053706146  8067491927
(2901)  8191197939  9520614196  6342875444  0643745123  7181921799
(2951)  9839101591  9561814675  1426912397  4894090718  6494231961

(3001)  5679452080  9514655022  5231603881  9301420937  6213785595
(3051)  6638937787  0830390697  9207734672  2182562599  6615014215
(3101)  0306803844  7734549202  6054146659  2520149744  2850732518
(3151)  6660021324  3408819071  0486331734  6496514539  0579626856
(3201)  1005508106  6587969981  6357473638  4052571459  1028970641
(3251)  4011097120  6280439039  7595156771  5770042033  7869936007
(3301)  2305587631  7635942187  3125147120  5329281918  2618612586
(3351)  7321579198  4148488291  6447060957  5270695722  0917567116
(3401)  7229109816  9091528017  3506712748  5832228718  3520935396
(3451)  5725121083  5791513698  8209144421  0067510334  6711031412

(3501)  6711136990  8658516398  3150197016  5151168517  1437657618
(3551)  3515565088  4909989859  9823873455  2833163550  7647918535
(3601)  8932261854  8963213293  3089857064  2046752590  7091548141
(3651)  6549859461  6371802709  8199430992  4488957571  2828905923
(3701)  2332609729  9712084433  5732654893  8239119325  9746366730
(3751)  5836041428  1388303203  8249037589  8524374417  0291327656
(3801)  1809377344  4030707469  2112019130  2033038019  7621101100
(3851)  4492932151  6084244485  9637669838  9522868478  3123552658
(3901)  2131449576  8572624334  4189303968  6426243410  7732269780
(3951)  2807318915  4411010446  8232527162  0105265227  2111660396

(4001)  6655730925  4711055785  3763466820  6531098965  2691862056
(4051)  4769312570  5863566201  8558100729  3606598764  8611791045
(4101)  3348850346  1136576867  5324944166  8039626579  7877185560
(4151)  8455296541  2665408530  6143444318  5867697514  5661406800
(4201)  7002378776  5913440171  2749470420  5622305389  9456131407
(4251)  1127000407  8547332699  3908145466  4645880797  2708266830
(4301)  6343285878  5698305235  8089330657  5740679545  7163775254
(4351)  2021149557  6158140025  0126228594  1302164715  5097925923
(4401)  0990796547  3761255176  5675135751  7829666454  7791745011
(4451)  2996148903  0463994713  2962107340  4375189573  5961458901

(4501)  9389713111  7904297828  5647503203  1986915140  2870808599
(4551)  0480109412  1472213179  4764777262  2414254854  5403321571
(4601)  8530614228  8137585043  0633217518  2979866223  7172159160
(4651)  7716692547  4873898665  4949450114  6540628433  6639379003
(4701)  9769265672  1463853067  3609657120  9180763832  7166416274
(4751)  8888007869  2560290228  4721040317  2118608204  1900042296
(4801)  6171196377  9213375751  1495950156  6049631862  9472654736
(4851)  4252308177  0367515906  7350235072  8354056704  0386743513
(4901)  6222247715  8915049530  9844489333  0963408780  7693259939
(4951)  7805419341  4473774418  4263129860  8099888687  4132604721
```

17.4 Digits 0 to 2,500 of π (base 16)

```
π  =  3.243F6A8885  A308D31319  8A2E037073  44A4093822  299F31D008
(0051)   2EFA98EC4E  6C89452821  E638D01377  BE5466CF34  E90C6CC0AC
(0101)   29B7C97C50  DD3F84D5B5  B547091792  16D5D98979  FB1BD1310B
(0151)   A698DFB5AC  2FFD72DBD0  1ADFB7B8E1  AFED6A267E  96BA7C9045
(0201)   F12C7F9924  A19947B391  6CF70801F2  E2858EFC16  636920D871
(0251)   574E69A458  FEA3F4933D  7E0D95748F  728EB65871  8BCD588215
(0301)   4AEE7B54A4  1DC25A59B5  9C30D5392A  F26013C5D1  B023286085
(0351)   F0CA417918  B8DB38EF8E  79DCB0603A  180E6C9E0E  8BB01E8A3E
(0401)   D71577C1BD  314B2778AF  2FDA55605C  60E65525F3  AA55AB9457
(0451)   48986263E8  144055CA39  6A2AAB10B6  B4CC5C3411  41E8CEA154

(0501)   86AF7C72E9  93B3EE1411  636FBC2A2B  A9C55D7418  31F6CE5C3E
(0551)   169B87931E  AFD6BA336C  24CF5C7A32  5381289586  773B8F4898
(0601)   6B4BB9AFC4  BFE81B6628  219361D809  CCFB21A991  487CAC605D
(0651)   EC8032EF84  5D5DE98575  B1DC262302  EB651B8823  893E81D396
(0701)   ACC50F6D6F  F383F44239  2E0B4482A4  84200469C8  F04A9E1F9B
(0751)   5E21C66842  F6E96C9A67  0C9C61ABD3  88F06A51A0  D2D8542F68
(0801)   960FA728AB  5133A36EEF  0B6C137A3B  E4BA3BF050  7EFB2A98A1
(0851)   F1651D39AF  017666CA59  3E82430E88  8CEE861945  6F9FB47D84
(0901)   A5C33B8B5E  BEE06F75D8  85C1207340  1A449F56C1  6AA64ED3AA
(0951)   62363F7706  1BFEDF7242  9B023D37D0  D724D00A12  48DB0FEAD3

(1001)   49F1C09B07  5372C98099  1B7B25D479  D8F6E8DEF7  E3FE501AB6
(1051)   794C3B976C  E0BD04C006  BAC1A94FB6  409F60C45E  5C9EC2196A
(1101)   246368FB6F  AF3E6C53B5  1339B2EB3B  52EC6F6DFC  511F9B3095
(1151)   2CCC814544  AF5EBD09BE  E3D004DE33  4AFD660F28  07192E4BB3
(1201)   C0CBA85745  C8740FD20B  5F39B9D3FB  DB5579C0BD  1A60320AD6
(1251)   A100C6402C  7279679F25  FEFB1FA3CC  8EA5E9F8DB  3222F83C75
(1301)   16DFFD616B  152F501EC8  AD0552AB32  3DB5FAFD23  876053317B
(1351)   483E00DF82  9E5C57BBCA  6F8CA01A87  562EDF1769  DBD542A8F6
(1401)   287EFFC3AC  6732C68C4F  5573695B27  B0BBCA58C8  E1FFA35DB8
(1451)   F011A010FA  3D98FD2183  B84AFCB56C  2DD1D35B9A  53E479B6F8

(1501)   4565D28E49  BC4BFB9790  E1DDF2DAA4  CB7E3362FB  1341CEE4C6
(1551)   E8EF20CADA  36774C01D0  7E9EFE2BF1  1FB495DBDA  4DAE909198
(1601)   EAAD8E716B  93D5A0D08E  D1D0AFC725  E08E3C5B2F  8E7594B78F
(1651)   F6E2FBF212  2B648888B8  12900DF01C  4FAD5EA068  8FC31CD1CF
(1701)   F191B3A8C1  AD2F2F2218  BE0E1777EA  752DFE8B02  1FA1E5A0CC
(1751)   0FB56F74E8  18ACF3D6CE  89E299B4A8  4FE0FD13E0  B77CC43B81
(1801)   D2ADA8D916  5FA2668095  770593CC73  14211A1477  E6AD206577
(1851)   B5FA86C754  42F5FB9D35  CFEBCDAF0C  7B3E89A0D6  411BD3AE1E
(1901)   7E4900250E  2D2071B35E  226800BB57  B8E0AF2464  369BF009B9
(1951)   1E5563911D  59DFA6AA78  C14389D95A  537F207D5B  A202E5B9C5

(2001)   8326037662  95CFA911C8  19684E734A  41B3472DCA  7B14A94A1B
(2051)   5100529A53  2915D60F57  3FBC9BC6E4  2B60A47681  E6740008BA
(2101)   6FB5571BE9  1FF296EC6B  2A0DD915B6  636521E7B9  F9B6FF3405
(2151)   2EC5855664  53B02D5DA9  9F8FA108BA  47996E8507  6A4B7A70E9
(2201)   B5B32944DB  75092EC419  2623AD6EA6  B049A7DF7D  9CEE60B88F
(2251)   EDB266ECAA  8C71699A17  FF5664526C  C2B19EE119  3602A57509
(2301)   4C29A05913  40E4183A3E  3F54989A5B  429D656B8F  E4D699F73F
(2351)   D6A1D29C07  EFE830F54D  2D38E6F025  5DC14CDD20  868470EB26
(2401)   6382E9C602  1ECC5E0968  6B3F3EBAEF  C93C971814  6B6A70A168
(2451)   7F358452A0  E286B79C53  05AA500737  3E07841C7F  DEAE5C8E7D
```

17.5 Digits 2,501 to 5,000 of π (base 16)

```
(2501)  44EC5716F2  B8B03ADA37  F0500C0DF0  1C1F040200  B3FFAE0CF5
(2551)  1A3CB574B2  25837A58DC  0921BDD191  13F97CA92F  F694324773
(2601)  22F547013A  E5E58137C2  DADCC8B576  349AF3DDA7  A94461460F
(2651)  D0030EECC8  C73EA4751E  41E238CD99  3BEA0E2F32  80BBA1183E
(2701)  B3314E548B  384F6DB908  6F420D03F6  0A04BF2CB8  129024977C
(2751)  795679B072  BCAF89AFDE  9A771FD993  0810B38BAE  12DCCF3F2E
(2801)  5512721F2E  6B7124501A  DDE69F84CD  877A584718  7408DA17BC
(2851)  9F9ABCE94B  7D8CEC7AEC  3ADB851DFA  63094366C4  64C3D2EF1C
(2901)  18473215D9  08DD433B37  24C2BA1612  A14D432A65  C451509400
(2951)  02133AE4DD  71DFF89E10  314E5581AC  77D65F1119  9B043556F1

(3001)  D7A3C76B3C  11183B5924  A509F28FE6  ED97F1FBFA  9EBABF2C1E
(3051)  153C6E86E3  4570EAE96F  B1860E5E0A  5A3E2AB377  1FE71C4E3D
(3101)  06FA2965DC  B999E71D0F  803E89D652  66C8252E4C  C9789C10B3
(3151)  6AC6150EBA  94E2EA78A5  FC3C531E0A  2DF4F2F74E  A7361D2B3D
(3201)  1939260F19  C279605223  A708F71312  B6EBADFE6E  EAC31F66E3
(3251)  BC4595A67B  C883B17F37  D1018CFF28  C332DDEFBE  6C5AA56558
(3301)  218568AB98  02EECEA50F  DB2F953B2A  EF7DAD5B6E  2F841521B6
(3351)  2829076170  ECDD477561  9F151013CC  A830EB61BD  960334FE1E
(3401)  AA0363CFB5  735C904C70  A239D59E9E  0BCBAADE14  EECC86BC60
(3451)  622CA79CAB  5CABB2F384  6E648B1EAF  19BDF0CAA0  2369B9655A

(3501)  BB5040685A  323C2AB4B3  319EE9D5C0  21B8F79B54  0B19875FA0
(3551)  9995F7997E  623D7DA8F8  37889A97E3  2D7711ED93  5F16681281
(3601)  0E358829C7  E61FD696DE  DFA17858BA  9957F584A5  1B2272639B
(3651)  83C3FF1AC2  4696CDB30A  EB532E3054  8FD948E46D  BC312858EB
(3701)  F2EF34C6FF  EAFE28ED61  EE7C3C735D  4A14D9E864  B7E342105D
(3751)  14203E13E0  45EEE2B6A3  AAABEADB6C  4F15FACB4F  D0C742F442
(3801)  EF6ABBB565  4F3B1D41CD  2105D81E79  9E86854DC7  E44B476A3D
(3851)  816250CF62  A1F25B8D26  46FC8883A0  C1C7B6A37F  1524C369CB
(3901)  749247848A  0B5692B285  095BBF00AD  19489D1462  B17423820E
(3951)  0058428D2A  0C55F5EA1D  ADF43E233F  70613372F0  928D937E41

(4001)  D65FECF16C  223BDB7CDE  3759CBEE74  604085F2A7  CE77326EA6
(4051)  07808419F8  509EE8EFD8  5561D99735  A969A7AAC5  0C06C25A04
(4101)  ABFC800BCA  DC9E447A2E  C3453484FD  D567050E1E  9EC9DB73DB
(4151)  D3105588CD  675FDA79E3  674340C5C4  3465713E38  D83D28F89E
(4201)  F16DFF2015  3E21E78FB0  3D4AE6E39F  2BDB83ADF7  E93D5A6894
(4251)  8140F7F64C  261C946929  34411520F7  7602D4F7BC  F46B2ED4A2
(4301)  0068D40824  713320F46A  43B7D4B750  0061AF1E39  F62E972445
(4351)  4614214F74  BF8B88404D  95FC1D96B5  91AF70F4DD  D366A02F45
(4401)  BFBC09EC03  BD97857FAC  6DD031CB85  0496EB27B3  55FD3941DA
(4451)  2547E6ABCA  0A9A285078  25530429F4  0A2C86DAE9  B66DFB68DC

(4501)  1462D74869  00680EC0A4  27A18DEE4F  3FFEA2E887  AD8CB58CE0
(4551)  067AF4D6B6  AACE1E7CD3  375FECCE78  A399406B2A  4220FE9E35
(4601)  D9F385B9EE  39D7AB3B12  4E8B1DC9FA  F74B6D1856  26A36631EA
(4651)  E397B23A6E  FA74DD5B43  326841E7F7  CA7820FBFB  0AF54ED8FE
(4701)  B397454056  ACBA489527  55533A3A20  838D87FE6B  A9B7D09695
(4751)  4B55A867BC  A1159A58CC  A9296399E1  DB33A62A4A  563F3125F9
(4801)  5EF47E1C90  29317CFDF8  E80204272F  7080BB155C  05282CE395
(4851)  C11548E4C6  6D2248C113  3FC70F86DC  07F9C9EE41  041F0F4047
(4901)  79A45D886E  17325F51EB  D59BC0D1F2  BCC18F4111  3564257B78
(4951)  34602A9C60  DFF8E8A31F  636C1B0E12  B4C202E132  9EAF664FD1
```

17.6 Continued fraction elements 0 to 1,000 of π

```
π = 3+[   7  15   1 292   1   1   1   2   1   3    1  14   2   1   1   2   2   2   2   1
(0021)   84   2   1   1  15   3  13   1   4   2    6   6  99   1   2   2   6   3   5   1
(0041)    1   6   8   1   7   1   2   3   7   1    2   1   1  12   1   1   1   3   1   1
(0061)    8   1   1   2   1   6   1   1   5   2    2   3   1   2   4   4  16   1 161  45
(0081)    1  22   1   2   2   1   4   1   2  24    1   2   1   3   1   2   1   1  10   2
(0101)    5   4   1   2   2   8   1   5   2   2   26   1   4   1   1   8   2  42   2   1
(0121)    7   3   3   1   1   7   2   4   9   7    2   3   1  57   1  18   1   9  19   1
(0141)    2  18   1   3   7  30   1   1   1   3    3   3   1   2   8   1   1   2   1  15
(0161)    1   2  13   1   2   1   4   1  12   1    1   3   3  28   1  10   3   2  20   1
(0181)    1   1   1   4   1   1   1   5   3   2    1   6   1   4   1 120   2   1   1   3

(0201)    1  23   1  15   1   3   7   1  16   1    2   1  21   2   1   1   2   9   1   6
(0221)    4 127  14   5   1   3  13   7   9   1    1   1   1   1   5   4   1   1   3   1
(0241)    1  29   3   1   1   2   2   1   3   1    1   1   3   1   1  10   3   1   3   1
(0261)    2   1  12   1   4   1   1   1   1   7    1   1   2   1  11   3   1   7   1   4
(0281)    1  48  16   1   4   5   2   1   1   4    3   1   2   3   1   2   2   1   2   5
(0301)   20   1   1   5   4   1 436   8   1   2    2   1   1   1   1   1   5   1   2   1
(0321)    3   6  11   4   3   1   1   1   2   5    4   6   9   1   5   1   5  15   1  11
(0341)   24   4   4   5   2   1   4   1   6   1    1   1   4   3   2   2   1   1   2   1
(0361)   58   5   1   2   1   2   1   1   2   2    7   1  15   1   4   8   1   1   4   2
(0381)    1   1   1   3   1   1   1   2   1   1    1   1   1   9   1   4   3  15   1   2

(0401)    1  13   1   1   1   3  24   1   2   4   10   5  12   3   3  21   1   2   1  34
(0421)    1   1   1   4  15   1   4  44   1   4 20776  1   1   1   1   1   1   1  23   1
(0441)    7   2   1  94  55   1   1   2   1   1    3   1   1  32   5   1  14   1   1   1
(0461)    1   1   3  50   2  16   5   1   2   1    4   6   3   1   3   3   1   2   2   2
(0481)    5   2   2   2  28   1   1  13   1   5   43   1   4   3   5   3   1   4   1   1
(0501)    2   2   1   1  19   2   7   1  72   3    1   2   3   7  11   1   2   1   1   2
(0521)    2   1   1   2   1   1   1   1   1  33    7  19   1  19   3   1   4   1   1   1
(0541)    1   2   3   1   3   2   2   2   2   4    1   1   1   4   2   3   1   1   1   1
(0561)   11   1   1   2   1   2   1   2   2   1    7   2  27   1   1   6   2   1   9   6
(0581)   26   1   1   3   2   1   1   1   1   1   15   1  36   4   2   2   1  22   2   1

(0601)  106   2   2   1   3   1  12  10   7   1    2   1   1   1   1   8   2   4   5   3
(0621)    2   1   4  23   1  18   2  10   3   1    6   6  13   8   6   2   2   2   2   1
(0641)    1   1   3   1   7  17   1   1   1   2    5   5   1   1   2  11   1   6   1   6
(0661)    1  29   4  29   3   5   3   1 141   1    2   7   7   2   2   7   1   1   7   1
(0681)    7   1   2   4   1   1   1  30   1  12    4  18  10   2   8   1   2   2   2   4
(0701)   13   1   5   4   1   6   1   1  11   2    4   2   1   1   3   3  12   1   1  39
(0721)    5   1   1  16 125   1   4   1   2   1   19   1   4   1   1   2   1   4   1  10
(0741)    1   4   2   1   1   1   5  10   4  14    1  13  41   1   4   1   8   1   1   2
(0761)    1   3   1   6   1   3   2   2   2   1    4   1  14   1   2   8   1   8   3   3
(0781)    3   1  37   4   2   4   1   3   4  25    4  27   2   7   1   1   2   6   1   1

(0801)    1  12   1   2   2   2  13  12   1   3    1   6   1   1  33   1   5   3   1   5
(0821)   15   8   8  47   1   3   2  12   2  12    1  12   1   2   5   3   1   1   1   1
(0841)    2   3   5   4   2   1   1   5   1   9   14   1   1   3   2   1   9   3  22  13
(0861)    1   1   3  20   1   1  61   1 376   2  107   1  10   3   2   2  31   1   2  10
(0881)    2   2  62   2   2   7   4   5   6   1    1   1   1   2   8   2  73   3   5  42
(0901)    1   3   2   1   1  59   6   1   1   1    5   1   6   1   2   6   1   1   1   1
(0921)    3   2   1   3   1   8   1   4   2   5    4   7   1   4   2   2   6   1   1   2
(0941)    2   1   1   1   1   1   2   1   2   2    5   1   2   1   1  10   1   6   1 129
(0961)    1   4  65   2   4   4   3   2   3   1    1   5   1   1   1   1   2   2   1
(0981)    2   1   1   2   2   1   2   3   1   2    1   2   4   2   1   2  27   6   2   1
```

17.7 Continued fraction elements 1,001 to 2,000 of π

```
(1001) 193   1    3   9    1   3  35   2   1    8   1   1   1   1    9   3  56    1    6   6
(1021)   2   8    1   8    1   2   3   6   3    1   3   1   1   1    2  13   1    1    1   1
(1041)  13   2    1   3    1   3  15   2   1    1   2   4   1   4    5   2   2    1    2   1
(1061)   6   1    4  12    1   1   1   1  13    1   3   4   1   1    1   2   9    1    7   1
(1081)   1   1    1   4    1   3   4   1   1    4   3   1  39   2    1   1   1    1    1   4
(1001)   7   2    2   2    1.  1   1   1   2  114  12   4   1   3    2   1  19    1    1   2
(1121)   1   1    3   4    1  60   3  72   2    1   1   1  50   1    1   1   1    3    1   1
(1141)   2   2    1   4    1   7   3   1   2    1   5   1   1   1    2   6   2   21    2   6
(1161)   1   6    1   1    2   1   7   1   8    1   1   5   4   1    1   1   1    1    1   1
(1181)   1   4    1  11    2   4  10   2   1    1  13   1   1   7   15   1   1    1    2   3
(1201)  15   8    8   2    1  13   3   5   1    2   1   6   1  10  123   3   1    4   59   4
(1221) 156  88    1   5    4   1   3   1   4    2   9   1   7   4    2   1   2    3    2   1
(1241)   2  11    1  13    7   7   1  63  37   12  86   1   1   1    1   2   2    4    2  18
(1261)   1   1    1  41    2   1   1  12   1    2   1   1   2  10    1   1   1    5    1   1
(1281)   3   1    7   5    1   9   1   2   2    7   1   1   5   2    1   3   3    5    2   1
(1301)  11   3    1   3    2   1   1   2   1   14   5   2   2   1    1   1   1    3    1   3
(1321)   3   2    2   1    3   2   1   2   1    4   1  14   1   1   58   7   1    2    1   1
(1341)   5   1    2   1    5  18   1   4   3    1   1   1   4   1    1   2   5    1  148   1
(1361)   9   2    1   2    1   5   4  93   1    1   2   4   1   2   73   1   1    3    1   1
(1381)   1   1    2   1   34   1   5   6   1    2   1   3   4   1   16  28  17    2    5   5
(1401)  26   1    1   4   12   1   3   2   1    5   1   2   9   3    2  41   1   16    2   2
(1421)  20   1   17   1    6  16   3   3   2    2   2  18  15   1    1  51   4    9    5   2
(1441)   2   1    2   1   45   3   1   1   3    1   2   1   3   1    1   3   5    1    2   3
(1461)   8   2   47   2    3   1   1   1  15    9   1   8   2   1    4   2   4   14    1  12
(1481)   2   1  161   1   26   2   1   2   1    1   1   1   2   2    1  18 528   12    4   1
(1501)   5  16    3   1    1   1   1   1   5    1   2   1  63   1   97   1   4    4   10   5
(1521)   9   5    2   3    2   5   7   1  32   13   1   5   4   1    7   1   3   12    1   3
(1541)   9   1    7   1  102  53   1   1   1    3   4   2  15   2    8   2   2    3    1   2
(1561)   4   1    1   3    2   3   1   1   2    3   1   1   6   1    1  14   1   80   11   1
(1581)   1   1    1  22    1   2   3   1   3   26   2  24   2   2    4   3   1    1    1   1
(1601)   3   1   63   1    1   1  25   1   1    1   8   1   3   3    1  10   5    6    2   1
(1621)   1   3    1   1    1   1   2   2   1    2   8  12   1  53    1   2   1    1    5   1
(1641)   1   3    1  39    1  12   1   3  14   18   9   3   2   2    2   1   1    3    1   4
(1661)   4   7    1  17    1  14   1   1   1    1   3   1   1  10    1   2   2    3    1   2
(1681)   1   2    2   2   12   1   3  44   2   10   1  14   1   2    1  43   4    1    7   3
(1701)   4   1    1   2    2   1  34   1   2    5   8   3   2   1    2  13   4    3    2   1
(1721)   1   1    1  25    1   5   1  94   2    4   3   4   5   1    1   1   1    1    1   2
(1741)   1   1    1   1    1   1   1   1   1   10  41   1   5   1    4   4   1  155    1   8
(1761)   1   1    1   1    4   1   1   2   9    2   1   2   1   1    1   6  23    1    2   3
(1781)   5   2    1   1    1   1   7  67   5    7   1  23   3   3    1   6   1   11    1  57
(1801)   1   4    1   5    1   1   8   1   1    2   5   2  10   1    1   2   1    1    3   1
(1821)   2   1    3   1   11   2  10   1   4   18   1   2   3   1    1   6   3    6    4  31
(1841)   3   4    1  18    3   9   7   5   1    2   2   1   7   1   23   2 217    1    2   1
(1861)   4   1   54   2  196  10   3   1  32    1  40  55   1   5    1   3   3    1    2   2
(1881)   1   3    6   3   16   1  31   1   5    6   1   4  42   4    1  10   1    3    1   3
(1901)   3   1    2   1    1   1   4   1  13    1  88   1   1   1   14   3  27    3    1   1
(1921)  16   4    1   2    4   1   4   1   1   17   2   4   1   1    9   2   1    1    3   1
(1941)   1  30    1   1    3   2   2   1   1    4  10   1   7   1    6   1  35    1    1   2
(1961)   3   6    1   1    2   4   4  24   1    1   1   1   1   1    3   1   1    2    1   2
(1981)   1   6    6   2    1   1  10   6   4    2   1   3   9   1    2  16   1    5    1   1
```

A. Documentation for the hfloat Library

The following is (almost) the documentation that comes with the hfloat *package*[1].

A.1 What hfloat is (good for)

- hfloat (for 'huge floats') is a library package for doing calculations with floating point numbers of extreme precision. It is optimised for computations with 1000...several million digits. The computations can be done in (almost) arbitrary radix.
- The library contains routines for addition, subtraction, multiplication, division, n-th power, square root, n-th root, logarithm, exponentiation and many more.
- There are implementations of several superlinear converging algorithms for the computation of pi=3.14159265... (in src/pi/). The computation of 1 million decimal digits of pi takes about 2 minutes on an AMD Athlon/800.
- Code examples for the usage of the library are in **examples/**.
- Code for a binary that collects the pi algorithms can be found in **calcpi/**.
- Included is the fxt-library, containing many FFT-implementations, code for convolution, correlation, spectrum and much more.
- High precision computations test your systems reliability (hardware and compiler): every little error results in garbage digits.
- Digits of appropriate constants can be used as high quality 'random' numbers eg. for cryptographic stuff.
- You may (and are encouraged to) use the fast multiplication in your own noncommercial bignum software.

---

[1] hfloat is online at **http://www.jjj.de/hfloat/**, the author of hfloat is Jörg Arndt.

A.2 Compiling the library

hfloat is developed under Linux with GNU C It has been compiled under Linux on intel (386, 486, 586 and compatibles and ia64), power pc, alpha and S/390. Compilation under other UNIXes should be possible with few or no changes, especially if GNU C is used. Basically you need a decent make-utility and a C-compiler[2], GNU C and GNU make are strongly recommended. To compile the library

- Type 'make dep' (for dependency files)
- Type 'make lib' to make the library

A.3 Functions of the hfloat library

Member functions of the hfloat class are declared in src/include/hfloat.h

Static member functions, they affect some property of the whole hfloat class:

- hfloat::default_prec() returns the default precision that can be used (unit is LIMBs)[3];
  hfloat::default_prec(unsigned long m) sets the default precision m;
  call it once at the beginning of your code
- hfloat::radix() returns the radix, which is the same for all hfloats;
  use hfloat::radix(unsigned long r) to set the radix.
  Note that after a radix change all hfloats contain garbage data in their mantissa so they must be reassigned new values before they are read.
  normally you will use hfloat::radix(unsigned long r) once at the beginning of your code

Nonstatic member functions, they affect one particular instance of an hfloat:

- the constructor hfloat() creates an hfloat with default_precision()

---
[2] expect problems with crippled Microsoft compiler-oids
[3] for the notion of a LIMB see page 251

A.3 Functions of the hfloat library

- the constructor hfloat(unsigned long n) creates an hfloat with n LIMBs
  Create hfloats like hfloat a; (for default precision) or hfloat a(1024); (for a precision of 1024 LIMBs).
  Direct initialization like hfloat a = 1234; is not possible because the constructor hfloat(unsigned long) has the explicit modifier. Instead say: hfloat a; a = 1234; .
- the copy constructor hfloat(const hfloat &h) creates a copy of the hfloat h
- the(assignment) operator =
  for arguments (i.e. right hand side quantities) int, long, unsigned long, double, strings (i.e. char *) and hfloat,
- size() returns the size of the mantissa (in LIMBs)
  size(unsigned long s) resizes the mantissa to s LIMBs
- prec() returns the current working precision (in LIMBs)
  prec(unsigned long p) sets the working precision to p LIMBs, resizes if necessary.
- exp() returns the exponent (with respect to LIMBs);
  exp(long) sets the exponent
- sign() returns $+1, 0, -1$ as usual;
  sign(int s) sets the sign

Operators for hfloats are declared in src/include/hfloatop.h:

- the shortcut operators +=, -=, *=, /=
  for right arguments long and hfloat,
- comparison operators ==, !=, >=, <=, <, >
  for comparisons of hfloats with hfloats or longs
- comparison operators <, >
  for comparisons of hfloats with doubles

Functions for hfloats are declared in src/include/hfloatfu.h:

- functions of the type func(src,result)
  where func $\in$ inv, sqr, sqrt, isqrt, cbrt, log[4], exp, ...
- functions of the type f(src1,src2,result)
  where func $\in$ add, sub, mul, div, pow, root, iroot, log, ...

The function names should be self-explanatory.
Print hfloat.h, hfloatfu.h and hfloatop.h now!

---

[4] logarithm to base 2.71828...

There are also the binary operators +, -, *, / and functions that return a `hfloat`, like `func(src)` where func ∈ inv, sqrt, ... but do not use them if you are after performance: they create temporary `hfloats` and are there only to allow lazy coding.

E.g. instead of
```
x = a + b;
```
use
```
x = a;   x += b;
```
or
```
add(a, b, x);
```
Instead of
```
x = exp(a);
```
use
```
exp(a,x);
```
Note that the result is always the rightmost argument.

A.4 Using hfloats in your own code

In order to write your own code that uses hfloats

- you must `#include src/include/hfloat.h` to get the functions of the hfloat lib.
- use `hfloat::default_prec(n)` where n is the precision in LIMBs (use a power of two)
- use `hfloat::radix(rx)` where rx is the radix (use 10000 for decimal numbers or 65536 for hex numbers)
- use `hfverbosity::tell_all()` if you like to have many operations echoed, `hfverbosity::hush_all()` for silent operations
- when compiling `yoursrc.cc` that uses `hfloats` link it against the library `libhflt.a`
- for extreme precisions increase the maximal workspace size, as described on page 250.

Cf. `examples/ex*.cc` for some simple examples of how to use hfloats. Look into `examples/index.txt` for what's there.

A.5 Computations with extreme precision

In order to do computations with the maximal possible precision on your computer you have to manually set the maximal workspace size:

Set the environment variable NOSWAP_BYTES according to the size of physical RAM where no swapping will occur, e.g. with bash say:
export NOSWAP_BYTES=32M
See src/fxt/auxil/workspace.cc for the default.

Do *not* give the total amount installed RAM, your machine will swap to death if you try really use it for hfloat. Currently if you don't give a power of 2 then the size is increased to the next bigger power of 2. You may append 'k' for kilobyte or 'M' for megabyte.

A.6 Precision and radix

For the quick readers I begin with the resumee:

- for decimal digits and precisions up to 2^{23} LIMBs[5] use radix 10,000 (for even greater precisions choose radix 1,000)
- for hexadecimal digits and precisions up to 2^{18} LIMBs[6] use radix 65,536 (for even greater precisions choose radix 4,096)

The mantissa of a **double** consists of (typically) 53 bits[7], its radix (base of representation) is 2. The analogous quantity in the mantissa of a hfloat is a LIMB (typedef'd to an unsigned 16 bit quantity). So the radix of a hfloat can be in the range $2...65536 (= 2^{16})$.

If one wants to get decimal numbers one would *not* use radix 10 but the greatest power of 10 that is $\leq 2^{16}$, which is 10000. Due to the implementation of the convolution there is a second restriction to the radix: The cumulative sums c_k have to be represented exactly enough to distinguish every (integer) quantity from the next bigger (or smaller) value. The highest possible value for a c_k, c_m, must not jump to $c_m \pm 1$ due to numerical errors. For radix R and a precision of N LIMBs

$$c_m = N(R-1)^2$$

This is N times the product of two 'nines' $(R-1)$, which can appear in the middle term of the convolution. c_m needs

$$\log_2(N(R-1)^2) = \log_2 N + 2\log_2(R-1)$$

---

[5] corresponding to more than 32 million decimal digits
[6] corresponding to more than 1 million hexadecimal digits
[7] Of which only the 52 least significant bits are physically present, the most significant bit is implied to be always set.

bits to be represented exactly.

Due to numerical noise there must be a few more bits for safety. The c_k are computed using doubles. We need to have

$$M \geq \log_2 N + 2\log_2(R-1) + S$$

where $S :=$ safetybits and $M :=$ mantissabits. With $\log_2(R-1) < \log_2(R)$ we have equivalently

$$N_{max}(R) = 2^{M-S-2\log_2(R)}$$

Thus if we want base 2 numbers we would first choose radix $R = 2^{16}$ but this can (if we have $M = 53$ mantissabits and require $S = 3$ safetybits) only be used for precisions up to

$$N_{max}(\text{radix} = 65,536) = 2^{53-3-2\cdot 16} = 2^{18} = 256 kilo\text{LIMBs}$$

(corresponding to $4096 kilo$ bits $= 1024 kilo$ hex digits).

$$N_{max}(\text{radix} = 32,768) = 2^{20} = 1 Mega\text{LIMBs}$$

(corresponding to $15360 kilo$ bits $= 3840 kilo$ hex digits).

$$N_{max}(\text{radix} = 16,384) = 2^{22} = 4 Mega\text{LIMBs}$$

(corresponding to $57344 kilo$ bits $= 14336 kilo$ hex digits).

For decimal numbers

$$N_{max}(\text{radix} = 10,000) = 2^{53-3-2\cdot 13.29} = 2^{23.42} > 8 Mega\text{LIMBs}$$

($N = 23$ corresponding to $32 Mega$ decimal digits).

$$N_{max}(\text{radix} = 1,000) = 2^{30.07} > 1 Giga\text{LIMBs}$$

($N = 30$ corresponding to $3 Giga$ decimal digits).

If the LIMBs weren't restricted to 16 bits:

$$N_{max}(\text{radix} = 100,000) = 2^{16.78} > 16 kilo\text{LIMBs}$$

($N = 16$ corresponding to $80 kilo$ decimal digits).

$$N_{max}(\text{radix} = 1,000,000) = 2^{10.13} > 1 kilo\text{LIMBs}$$

($N = 10$ corresponding to $6 kilo$ decimal digits).

A.7 Compiling & running the π-example-code

To compile the code in the directory `calcpi/`

- Type 'make pi'.
- cd to the `bin` directory, find the example binary `pi` there.
- Type 'pi --help' for usage information (the help text can also be found in the file `doc/pihelp.txt`).
- Type 'pi 10' to run the program. It will compute 4096 decimal digits of π (in about 1 second on a not very ancient computer).
- See the file `result.txt` for the digits of π. All but the last few digits (10 or so) of the output should be correct. Check correctness like this:
- Type 'pi 10 314' to run the program selftest which uses several algorithms and outputs how many LIMBs differ in the results.
- Type 'pi 8 0 65536' to get 1024 hexadecimal digits of π.
- Type 'pi 18 16' to get > 1 million decimal digits of π.

A.8 Structure of hfloat

The hfloat code is divided into two layers:

The class hfloat (top, user interface) layer. A hfloat consists of a unique id number (for internal use only), sign and exponent plus and a pointer to class hfdata. Currently there is a one-to-one correspondence between hfloats and hfdata (it possibly will remain like this). The arithmetical operators and functions like `mul(src1,src2,dest)` are here. The iterations (for root extraction, inversion, ...) are also in this layer.
Declarations are in `src/include/hfloat.h` (class), `src/include/hfloatop.h` (operators) and `src/include/hfloatfu.h` (functions). Source files are in `src/hf/`.

The class hfdata (mantissa operations, number crunching) layer. A hfdata consists of a pointer to the digit field, the size and precision of the mantissa data. Typical function names are `dt_something(...)`, Declarations are in `src/include/hfdata.h` (class) and `src/include/hfdatafu.h` (functions). Source files are in `src/dt/`.

A.9 Organisation of the files

directory	which files are there
bin/	binaries and libs
doc/	the main documentation
examples/	simple example code, look there!
calcpi/	π example
src/	(source directory)
src/fxt/	source for the fxt-librarry
src/fxt/mult/	source for the fft-multiplication
src/include/	header files for the hfloat-librarry
src/tz/	source for the transcendental functions
src/hf/	source for the hfloat-layer
src/dt/	source for the hfdata-layer
src/pi/	source for the π-library
testing/	cryptic test routines, better ignore
*/bucket/	garbage & trash

Selected files in some directories:

- src/include/: (declarations)
  - hfloat.h: the hfloat class
  - hfloatfu.h: hfloat functions
  - hfloatop.h: hfloat operators
  - hfverbosity.h: adjust how talkative the hfloat operations are
  - hfdata.h: the hfdata class
  - hfdatafu.h: hfdata functions
- src/pi/: implementations of the π-algorithms
- src/hf/:
  - hfloat.cc: hfloat class member functions
  - hfloatop.cc: hfloat class operators
  - init.cc: magic initialiser for the hfloat class
  - it*.cc: code for iterations (inverse, sqrt, inverse n-th root)
- src/dt/:
  - hfdata.cc: hfdata class member functions
  - dt*.cc: hfdata functions

A.10 Distribution policy & no warranty

———— legal stuff ————

- The hfloat package and code is freeware, you may use it at no cost for noncommercial purposes.
- The copyright remains by the author (Jörg Arndt, arndt@jjj.de). Obvious exceptions are the included pieces of code by other authors.
- You may (and are encouraged to) give hfloat away free of charge. Always give a pointer to the original hfloat if you redistribute pieces of it.
- You are not allowed to make money with hfloat (or parts of it) in any way.
- Before putting hfloat on a CD or similar get my agreement.
- WARNING: hfloat is distributed 'as is', you are using hfloat at your own risk. I will take no responsibility for potential damage that might be caused.

———— end of legal stuff ————

- Please be nice and give me credits if you use hfloat for anything interesting/noticable/scientific. I am interested in hearing about your project.
- If you use hfloat for scientific purposes then cite it in your publications as you would cite an ordinary publication.
- Please check for newer versions before passing hfloat on. I tend to upload new versions even for small changes.
- If you notice errors in the code or the documentation please take the time and drop me an email.

Bibliography

1. Victor Adamchik, Stan Wagon, *A Simple Formula for π*, American Mathematical Monthly, Vol 104 (1997) 852–855, also in [20, pp. 557–559].
   (Cited on p. 126)
2. Victor Adamchik, Stan Wagon, *Pi: A 2000-Year Search Changes Direction*, Education and Research, Vol 5 (1996), No. 1, 11–19,
   online at http://members.wri.com/victor/articles/pi.html
   (Cited on p. 19, 126, 227)
3. Association pour le Devéloppement de la Culture Scientifique (ADCS), 61 rue Saint-Fuscien, 8000 Amiens France, *Le nombre π*, Sonderheft zur Zeitschrift *Le PETIT ARCHIMÈDE*.
4. Timm Ahrendt, *Schnelle Berechnung der komplexen Quadratwurzel auf hohe Genauigkeit*, Logos Verlag, Berlin, 1996.
   Online at http://web.informatik.uni-bonn.de/II/staff/ahrendt/AHRENDTliteratur.html
   (Cited on p. 148)
5. Gert Almquist, *Many Correct Digits of π, Revisited*, American Mathematical Monthly, Vol. 104 (1997), No. 4, 351–353.
   (Cited on p. 157)
6. Archimedes, *Measurement of a Circle*, in *The Works of Archimedes*, Ed. by T.L. Heath, Cambridge University Press, 1897, also in [20, pp. 7–14].
   (Cited on p. 171)
7. Jörg Arndt, *Remarks on arithmetical algorithms and the compuation of π*, Online at http://www.jjj.de/joerg.html.
   (Cited on p. 75, 110, 233, 234)
8. Nigel Backhouse, *Pancake functions and approximations to π*. The Mathematical Gazette, Vol. 79 (1995), 371–374.
   (Cited on p. 234)
9. David H. Bailey, *The Computaion of π to 29,360,999 Decimal Digits Using Borweins' Quartically Convergent Algorithm*, Mathematics of Computation, Vol. 50, No. 181 (Jan. 1988), 283–296, also in [20, pp. 562–575].
   (Cited on p. 202)
10. David Bailey, Peter Borwein and Simon Plouffe, *On The Rapid Computation of Various Polylogarithmic Constants*, Online at http://www.cecm.sfu.ca/~pborwein, also in [20, pp. 663–676].
    (Cited on p. 118, 123)
11. David H. Bailey, Jonathan M. Borwein and Peter B. Borwein, *Ramanujan, Modular Equations, and Approximations to Pi or How to compute One Billion Digits of Pi*, American Mathematical Monthly, Vol. 96 (1989), 201–219, also in [20, pp. 623–641].
    (Cited on p. 6, 105)

12. David H. Bailey, Jonathan M. Borwein and Richard E. Crandall, *On the Khintchine Constant*, Mathematics of Computation, Vol. 66, No. 217 (Jan. 1997), 417–431.
    (Cited on p. 68)
13. David H. Bailey, Jonathan M. Borwein, Peter B. Borwein and Simon Plouffe, *The Quest for Pi*, The Mathematical Intelligencer, Vol. 19 (1997), No. 1, 50–56.
    (Cited on p. 17, 119, 122, 197, 200, 237, 238)
14. Walter William Rouse Ball, *Mathematical Recreations and Essays*, 11th edition with revisions, Macmillan & Co Ltd, London, 1963.
    (Cited on p. 41, 180, 181, 182, 183)
15. Walter William Rouse Ball, Harold Scott Macdonald Coxeter, *Mathematical Recreations and Essays*, Toronto, 1974.
    (Cited on p. 25)
16. J. P. Ballantine, *The best (?) formula for computing π to a thousand places*, American Mathematical Monthly, Vol. 46 (1939), 499–501.
    (Cited on p. 74)
17. Friedrich L. Bauer, *Decrypted Secrets*, Methods and Maxims of Cryptology, Second, Revised and Extended Edition, Springer-Verlag, Heidelberg, 2000.
    (Cited on p. 61, 228)
18. Petr Beckmann, *A History of π (PI)*, Fifth Edition, The Golem Press, Boulder, Colorado, 1982.
    (Cited on p. 64, 188, 194, 195)
19. Fabrice Bellard, *Computation of the n'th digit of pi on any base in $O(n^2)$*, Online at http://www-stud.enst.fr/~bellard/pi/.
    (Cited on p. 128, 227, 229, 234)
20. Lennart Berggren, Jonathan Borwein, Peter Borwein, *Pi: A Source Book*, Springer-Verlag, New York, 1997.
    (Cited on p. 115, 187, 188, 226)
21. Bruce C. Berndt, *Ramanujan—100 Years Old (Fashioned) or 100 Years New (Fangled)?*, The Mathematical Intelligencer, Vol. 10 (1988), No. 3, 24–29.
    (Cited on p. 105)
22. Bruce C. Berndt, *A Pilgrimage*, The Mathematical Intelligencer, Vol. 8 (1986), No. 1, 25–30.
    (Cited on p. 105, 106)
23. Bruce C. Berndt and Robert A. Rankin, *Ramanujan, Letters and Commentary*, American Mathematical Society, London Mathematical Society, 1995.
    (Cited on p. 105)
24. Bruce C. Berndt, Ramanujan's Notebooks, Part I(1985), II(1989), III(1991), IV(1994), V(to appear), Springer-Verlag, New York.
    (Cited on p. 109)
25. Bruce C. Berndt and S. Bhargava, *Ramanujan – For Lowbrows*, American Mathematical Monthly, Vol. 100 (1993), 644–656.
    (Cited on p. 58, 108)
26. Eugen Beutel, *Die Quadratur des Kreises*, Mathematische Bibliothek, No. 12, Verlag von B.G. Teubner, Leipzig und Berlin, 1913.
    (Cited on p. 181)
27. Ludwig Bieberbach, *Persönlichkeitsstruktur und mathematisches Schaffen*, Forschungen und Fortschritte, 10. Jahrg., Nr. 18 (20. Juni 1934), 235–237.
28. David Blatner, *The Joy of π*, Walker and Company, New York, 1997.
    (Cited on p. 201)

29. Jonathan M. Borwein, *Brouwer-Heyting Sequences Converge*, Mathematical Intelligencer, Vol. 20, No. 1, 1998, 14–15.
    (Cited on p. 30)
30. Jonathan M. Borwein and Peter B. Borwein, *The Arithmetic-Geometric Mean and Fast Computations of Elementary Functions*, SIAM Review, Vol. 26, No. 3 (July 1984), 351–366, also in [20, pp. 537–552].
    (Cited on p. 114, 236)
31. Jonathan M. Borwein, Peter B. Borwein, *More Ramanujan-type Series for $1/\pi$*, Proceedings of the Centenary Conference, Univ. of Illinois at Urbana-Champaign, June 1-5, 1987, 359–374.
32. Jonathan M. Borwein, Peter B. Borwein, *Pi and the AGM – A Study in Analytic Number Theory and Computational Complexity*, Wiley, N.Y., 1987.
    (Cited on p. 7, 59, 73, 74, 90, 92, 114, 188, 228, 229, 237)
33. Jonathan M. Borwein, Peter B. Borwein, *Ramanujan and Pi*, Science and Applications; Supercomputing 88: Volume II, 117-128, 1988. also in [20, p. 588–595]
    (Cited on p. 105, 107, 111, 237)
34. Jonathan M. Borwein, Peter B. Borwein, *Strange Series and High Precision Fraud*, American Mathematical Monthly, Vol. 99 (1992), 622–640.
    (Cited on p. 63, 230)
35. Jonathan M. Borwein, Peter B. Borwein and K. Dilcher, *Pi, Euler Numbers, and Asymptotic Expansions*, American Mathematical Monthly, Vol. 96 (1989), 681–687, also in [20, pp. 642–648].
    (Cited on p. 156, 157)
36. Jonathan M. Borwein and Roland Girgensohn, *Addition theorems and binary expansions*, Canadian Journal of Mathematics, Vol. 47 (1995), 262–273.
    (Cited on p. 48, 235)
37. Richard P. Brent, *Fast Multiple-Precision Evaluation of Elementary Functions*, Journal of the ACM, Vol. 23, No. 2, April 1976, 242–251, also in [20, pp. 424–433].
    (Cited on p. 87)
38. E. Oran Brigham, *The Fast Fourier Transform*, Prentice Hall, 1994.
    (Cited on p. 199)
39. David M. Burton, *Burton's History of Mathematics: An Introduction*, Wm. C. Brown Publishers, Third Edition, 1995.
    (Cited on p. 14, 190)
40. George S. Carr, *Formulas and Theorems in Pure Mathematics*, Second Edition, Chelsea Publishing Company, 1970.
    (Cited on p. 106)
41. Dario Castellanos, *The Ubiquitous π*, Mathematical Magazine 61 (1988), 67–98 (Part I) and 148–163 (Part II).
    (Cited on p. 45, 59, 60, 72, 154, 166, 180, 190, 198, 224, 225, 226, 227, 228, 229, 232)
42. Henri Cohen, F. Rodriguez Villegas and Don Zagier *Convergence acceleration of alternating series*, Experimental Math., to appear. downloadable from http://www.math.u-bordeaux.fr/ cohen/
    (Cited on p. 150)
43. J.W. Cooley, and J.W.Tukey, *An algorithm for the machine calculation of complex Fourier series*, Mathematics of Computation Vol. 19 (1965), No. 90, 297–301.
    (Cited on p. 137)

44. James W. Cooley, Peter A.W. Lewis and Peter D. Welch, *Historical Notes on the Fast Fourier Transform*, Proceedings of the IEEE, Vol. 55 (October 1967), No. 10, 1675–1677.
    (Cited on p. 199)
45. David A. Cox, *The arithmetic-geometric mean of Gauss*, L'Enseignment Mathématique, t. 30 (1984), 275–330, also in [20, pp. 481–536].
    (Cited on p. 95, 96, 97, 98)
46. David A. Cox, *Gauss and the Arithmetic-Geometric Mean*, Notices of the American Mathematical Society 32 (1985), 147–151.
    (Cited on p. 95)
47. Richard E. Crandall, *Projects in Scientific Computation*, Springer-Verlag, New York, Inc., 1994.
    (Cited on p. 230)
48. Jean-Paul Delahaye, *Pi – die Story*, Birkhäuser-Verlag, Basel, 1999. Französische Originalausgabe: *Le fascinant nombre π*, Pour La Science, 1997.
    (Cited on p. 50)
49. Drinfel'd, *Quadratur des Kreises und Transzendenz von π*, VEB Deutscher Verlag der Wissenschaften, Berlin, 1980.
    (Cited on p. 197)
50. Underwood Dudley, *Mathematical Cranks*, The Mathematical Association of America, 1992.
    (Cited on p. 8)
51. Heinz-Dieter Ebbinghaus, Reinhold Remmert, et al., *Numbers*, Springer-Verlag, New York, 1990.
    (Cited on p. 56, 168, 170, 176, 213)
52. Leonhard Euler, *Introduction to Analysis of the Infinite*, Translation of: Introductio in analysin infinitorum, Book I, Translated by John D. Blanton, Springer-Verlag, New York, 1988, Chapter 10 also in [20, pp. 112–128].
    (Cited on p. 65, 166, 194)
53. Leonhard Euler, *Einleitung in die Analysis des Unendlichen*, Erster Teil. Ins Deutsche übertragen von H. Maser, Berlin. Verlag von Julius Springer. 1885.
    (Cited on p. 166, 194)
54. Martin Gardner, *Some comments by Dr. Matrix on symmetries and reversals*, Scientific American, January 1965, 110–116.
    (Cited on p. 24)
55. Carl Friedrich Gauß, *Mathematisches Tagebuch, 1796–1814*, Reihe Ostwalds Klassiker der exakten Wissenschaften, Akademische Verlagsgesellschaft Geest & Portig K.-G., Leipzig 1985. Also in [56, X, pp. 483–574].
    (Cited on p. 98)
56. Carl Friedrich Gauß, *Werke*, Göttingen, 1866–1933.
    (Cited on p. 73, 88, 95, 99, 232)
57. C.F. Gauß, *Nachlass zur Theorie des Arithmetisch-Geometrischen Mittels und der Modulfunktion*, herausgegeben von Harald Geppert, Reihe Ostwald's Klassiker der exakten Wissenschaften, Nr. 225, Akademische Verlagsgesellschaft, Leipzig, 1927.
    (Cited on p. 95)
58. Roland Girgensohn, personal E-Mail, 1997.
    (Cited on p. 119)
59. Franz Gnädinger, *Primary Hill And Rising Sun*, Online at http://www.access.ch/circle/.
    (Cited on p. 210)

60. Wei Gong-yi, Yang Zi-qiang, Sun Jia-chang, Li Lia-kai, *The Computation of π to 10,000,000 Decimal Digits* J. Num. Method & Comp. Appl., 17:1(1996), 78–81.
   (Cited on p. 158)
61. Groß, Zagier, *On singular moduli*, J. Reine Angew. Math., 355(1985), 191–220.
   (Cited on p. 27)
62. Bruno Haible und Thomas Papanikolaou, *Fast multipliprecision evaluation of series of rational numbers*, Online at
   http://www.informatik.th-darmstadt.de/TI/Veroeffentlichung/TR/Welcome.html.
   (Cited on p. 215, 217)
63. Godfrey H. Hardy, *Ramanujan, Twelve Lectures on Subjects Suggested by his Life and Work*, Chelsea Publishing Company, New York, 1940.
   (Cited on p. 106, 108, 109)
64. Sir Thomas Heath, *A History of Greek Mathematics*, Volume II From Aristarchus To Diophantus, Dover Publications, 1981, first published 1921.
   (Cited on p. 173, 174)
65. M. Heidemann, D. Johnson, C. Burrus, *Gauss and the history of the Fast Fourier transformation*, IEEE ASSP Magazine 1, 1984, 14–21.
   (Cited on p. 137, 199)
66. Joseph E. Hofmann, *Geschichte der Mathematik*, Sammlung Göschen Band 226 (Erster Teil), Band 875 (Zweiter Teil), Band 882 (Dritter Teil), Walter de Gruyter & Co, Berlin 1953.
67. Dirk Huylebrouck, *The π-Room in Paris*, The Mathematical Intelligencer, Vol. 18, No. 2, 1996, 51–53.
   (Cited on p. 50)
68. William Jones, *Synopsis Palmariorum Mathesos*, London, 1706, In extracts in [20, pp. 108–109].
69. A.P. Juschkewitsch, *Geschichte der Mathematik im Mittelalter*, Teubner Verlag, 1964.
   (Cited on p. 182)
70. Sven Kabus, *Untersuchung verschiedener π-Algorithmen am PC*, (PC investigation of various π algorithms), entry in the *Jugend forscht* (Youth researches) competition, 1998, Junge wissenschaft (Friedrich-Verlag), 58 (1999), 34–42.
   (Cited on p. 63)
71. Robert Kanigel, *The Man Who Know Infinity*, Macmillan Publishing Company, 1991.
   (Cited on p. 105)
72. Victor J. Katz, *A History of Mathematics*, HarperCollins College Publishers, 1993.
   (Cited on p. 167, 168, 179)
73. Theo Kempermann, *Zahlentheoretische Kostproben*. Verlag Harri Deutsch, Thun, Frankfurt am Main, 1995.
   (Cited on p. 14, 158, 234)
74. Alexander Khintchine, *Continued Fractions*, Nordhoff, Groningen 1963 [Moskow 1935].
   (Cited on p. 68)
75. Louis V. King, *On the Direct Numerical Calculation of Elliptic Functions and Integrals*, Cambridge University Press, Cambridge, 1924.
   (Cited on p. 57)

76. Konrad Knopp, *Infinite sequences and series* Dover Publications, New York, 1956.
    (Cited on p. 71, 78)
77. Donald E. Knuth, *The Art of Computer Programming*, Volume 2 *Seminumerical Algorithms*, Third Edition, 1998, 4th Printing August 1999.
    (Cited on p. 28, 102, 120, 121, 132, 135, 180, 182)
78. Lam Lay-Yong and Ang Tian-Se, *Circle Measurement in Acient China*, Historia Mathematica, Vol. 13 (1986), 325–340, also in [20, pp. 20–44].
    (Cited on p. 176)
79. Johann Heinrich Lambert, *Mémoire sur quelques propriétés remarquables des quantités transcendantes circulaires et logarithmiques*, Memoires de l'Academémie de sciences de Berlin, [17] (1761), 1768, 265–322, also in [20, pp. 129–140].
    (Cited on p. 192)
80. L.J. Lange *An elegant new continued fraction for π*, American Mathematical Monthly, Vol 106 (May 1999) 456–458.
    (Cited on p. 33, 231)
81. Nick Lord, *Recent calculations of π: the Gauss-Salamin algorithm*, The Mathematical Gazette, Vol. 76 (1992), 231–242.
    (Cited on p. 100)
82. Philipp Maennchen, *Gauß als Zahlenrechner*, in [56, X.2,1-73].
    (Cited on p. 57)
83. Eli Maor, *e: the story of a number*, Princeton University Press, 1994.
    (Cited on p. 96)
84. Yuri V. Matiyasevich, Richard K. Guy, *A New Formula for π*, American Mathematical Monthly, Vol. 93, (October 1986), 631–635.
    (Cited on p. 234)
85. Niels Nielsen, *Die Gammafunktion*, Chelsea Publishing Company, Bronx, New York, 1965 (first published 1906).
    (Cited on p. 229)
86. Ivan Niven, *Irrational Numbers*, John Wiley and Sons, 1956.
    (Cited on p. 22)
87. C.D. Olds, *Continued Fractions*, Mathematical Association Of America, 1963.
    (Cited on p. 65)
88. Thomas J. Osler, *The Union of Vieta's and Wallis's Products for Pi*, American Mathematical Monthly, Vol. 106, (October 1999), 774–776.
89. S. Parameswaran, *Whish's showroom revisited*. The Mathematical Gazette, 76(1992), 28–36.
    (Cited on p. 186, 223, 224)
90. Oskar Perron, *Die Lehre von den Kettenbrüchen*, Band I und II, B.G. Teubner Verlagsgesellschaft, Stuttgart, 3. Auflage, 1957.
    (Cited on p. 65, 67)
91. G.M. Phillips, *Archimedes and the numerical analyst*, American Mathematical Monthly, Vol. 88(1981), 165–169, also in [20, pp. 15–19].
92. Richard Preston, *The Mountains of pi*, The New Yorker, Mar. 2, 1992, 36–67.
    (Cited on p. 2, 10, 34, 201)
93. Martin R. Powell, *Significant insignificant digits*, The Mathematical Gazette, 66(1982), 220–221.
    (Cited on p. 156)
94. Stanley Rabinowitz, *Abstract 863-11-482: A Spigot Algorithm for Pi*, Abstracts Amer. Math. Society, 12(1991), 30.
    (Cited on p. 37)

95. Stanley Rabinowitz and Stanley Wagon, *A Spigot Algorithm for the Digits of Pi*, American Mathematical Monthly, Vol. 103 (March 1995), 195–203.
    (Cited on p. 37, 77, 82)
96. Srinivasa Ramanujan, *Modular equations and approximations to π*, Quart. J. Math. 45 (1914) 350–372, also in [20, pp. 241–257].
    (Cited on p. 27, 57, 58, 103, 105, 226)
97. Srinivasa Ramanujan, *Squaring the Circle*, J. Indian Math. Soc. 5(1913), 132, also in [20, p. 240].
    (Cited on p. 58)
98. Ranjan Roy, *The Discovery of the Series Formula for π by Leibnitz, Gregory and Nilakantha*, Mathematics Magazine 63 (1990), 291–306, also in [20, pp. 92–107].
    (Cited on p. 186, 230)
99. Hansklaus Rummler, *Squaring the Circle with Holes*, American Mathematical Monthly, Vol. 100 (November 1993), 858–860.
    (Cited on p. 162)
100. Eugene Salamin, *Computation of π Using Arithmetic-Geometric Mean*, Mathematics of Computation, Vol. 30, No. 135, July 1976, 565–570, also in [20, pp. 418–423].
    (Cited on p. 87, 102, 231)
101. Eugene Salamin, E-Mail to Jörg Arndt, 1997.
    (Cited on p. 102)
102. Norbert Schappacher, *Edmund Landau's Göttingen: From the Life and Death of a Great Mathematical Center*, The Mathematical Intelligencer, Vol. 13, No. 4, 12–18.
103. L. Schlesinger, *Über Gauss Funktionentheoretische Arbeiten*, in [56, X.2, 9–90].
104. Arnold Schönhage und Volker Strassen, *Schnelle Multiplikation großer Zahlen*, Computing 7 (1971), 281–292.
    (Cited on p. 141, 147, 217)
105. Arnold Schönhage, A. F. W. Grotefeld/E. Vetter, *Fast Algorithms, A Multitape Turing Machine Implementation*, BI Wissenschaftsverlag, Mannheim, 1994.
    (Cited on p. 92, 236)
106. D. Shanks and J. W. Wrench, *Calculation of Pi to 100,000 Decimals*, Mathematics of Computation, vol. 16 (1962), 76–79, also in [20, p. 326–329].
    (Cited on p. 49, 197)
107. Daniel Shanks, *Dihedral Quartic Approximations and Series for π*, Journal of Number Theory 14(1982), 397–423.
    (Cited on p. 62)
108. Daniel Shanks, *Improving an Approximation for Pi*, American Mathematical Monthly, Vol. 99 (1992) No. 3, 263.
    (Cited on p. 49)
109. Allen Shields, *Klein and Bieberbach, Mathematics, Race, and Biology*, The Mathematical Intelligencer, Vol. 10 (1988), No. 3, 7–11.
    (Cited on p. 214)
110. David Singmaster, *The legal values of π*, The Mathematical Intelligencer, Vol. 7 (1985), No. 2, 69–72, also in [20, pp. 236–239].
    (Cited on p. 213)
111. M.D. Stern, *A Remarkable Approximation to π*, The Mathematical Gazette, Vol. 69 (1985), 218–219, also in [20, pp. 460–461].
    (Cited on p. 169)

112. Lutz von Strassnitzky, *Der Kreis-Umfang für den Durchmesser 1 auf 200 Decimalstellen berechnet* von Herrn Z. Dahse in Wien. Journal für die reine und angewandte Mathematik, Bd 27, 1844, 198.
(Cited on p. 194)

113. D. Takahasi, Y. Kanada, *Calculation of Pi to 51.5 Billion Decimal Digits on Distributed Memory and Parallel Processors*, Transactions of Information Processing Society of Japan, Vol. 39, No. 7, (1998).
(Cited on p. 201, 206)

114. Johannes Tropfke, *Geschichte der Elementarmathematik*, Vierter Band, Ebene Geometrie, Dritte Auflage, Walter De Gruyter & Co, 1940.
(Cited on p. 73, 165, 166, 169, 171, 175, 176, 182, 183, 187, 189, 224, 228, 231, 232, 235)

115. Franciscus Vieta (François Viète), *Variorum de Rebus Mathematicis*, 1593, Caput XVIII, 400, auch (tw.) in [20, p. 53–56].
(Cited on p. 187)

116. Alexei Volkov, *Zhao Youquin and His Calculation of π*, Historia Mathematica, Vol. 24, No. 3, August 1997, 301–331.
(Cited on p. 179)

117. Stan Wagon, *Is π Normal?* The Mathematical Intelligencer, Vol. 7 (1985), No. 3, 65–67, also in [20, pp. 557–559].
(Cited on p. 23)

118. Sebastian Wedeniwski, *Piologie, Eine exakte arithmetische Bibliothek in C++*, Edition 1.3, http://www.hipilib.de/
(Cited on p. 234)

119. David Wells, *The Penguin Dictionary of Curious and Interesting Numbers*, Penguin Books, Middlesex, 1986.
(Cited on p. 195)

120. David Wells, *Are These the Most Beautiful ?* The Mathematical Intelligencer, Vol. 12, No. 3, 37–41.
(Cited on p. 190)

121. Jet Wimp, *Book Review of Pi and the AGM* in SIAM Review, Oct. 1988, 530–533.
(Cited on p. 116)

122. J. W. Wrench, Jr., *The evolution of extended decimal approximations to π*, The Mathematics Teacher, vol. 53 (Dec. 1960), 544–650. also in [20, p. 319–325].
(Cited on p. 73, 196)

123. Robert M. Young, *Probability, pi, and the primes, Serendipity and experimentation in elementary calculus*, The Mathematical Gazette, Vol. 82 (Nov. 1998), 443–451.
(Cited on p. 9)

Index

163 phenomenon, 25

Abel, 95, 98, 109
ADA, 118
Ada of Lovelace, 118
Adamchik, 19, 126, 203, 227
AGM, 87–102
– basic features, 90
– rule, 88
Aitken, 25
Al-Khashî, 182, 205
algebra, 180
algorithm, 180
– Archimedes, 235
– BBP, 119
– binsplit, 215, 216, 218
– Borwein
– – cubic, 237
– – nonic, 238
– – quadratic, 114, 236
– – quartic, 114, 237
– – quintic, 111, 237
– dartboard, 39
– Descartes, 235
– Gauss AGM, 77, 90, 91, 93, 114, 119, 200, 236
– Kabus, 64
– spigot, 77
– sumalt, 150
Alkarism, 180, 205
Andersson, 153
Andrews, 109
Antiphon, 170
approximation
– Archimedes, 52, 171
– Borweins, 59, 62
– Castellanos, 59
– Plouffe, 60
– Ramanujan, 57, 58
– Shanks, 62
– Tsu Chhung-Chih, 52, 179
approximations for π, 51–64

Archimedes, 15, 69, 170–177, 205, 235
arctan relations, 232–234
Arcus Tangens, 69–76
Aristophanes, 7
arithmetic, 131–152
arithmetic-geometric mean see AGM 87
Arndt, 75, 76, 110, 233, 234
Aryabhata, 56, 179, 205
Au-Yeung, 235

Babylon, 69, 167, 205
Backhouse, 234
Bailey, 19, 117, 123, 202, 206, 207, 227, 235
Ball, 41
Barrow, 166
Bauer, 55, 61, 228
Bayes, 21
BBP algorithm, 117–129
BBP series, 118, 119
BBP-like series, 227–228
Beckmann, 188, 191, 194, 195
Bellard, 127, 128, 207, 227, 229, 234
Berggren, 115
Berndt, 105, 109
Bernoulli, Jakob, 96, 191
Bernoulli, Johann, 96, 166, 191
"best (arctan) formula", 74
Bible, 4, 51, 169, 205
Bieberbach, 213
binary modulo exponentiation, 120
binsplit algorithm, 215, 216, 218
bit recursion, 48, 235
Blatner, 201
Borwein, J., 30, 48
Borwein, J. and P., 11, 17, 59, 63, 111, 113, 114, 157, 200, 229, 230, 236–238
– algorithm
– – cubic, 237
– – nonic, 238
– – quadratic, 236

-- quartic, 114, 237
-- quintic, 111, 237
Borwein, P., 19, 117, 198, 207, 227
Borweins and π, 113–116
Boyer, 206
Brahmagupta, 180, 205
Brent, 87, 200, 236
Brent-Salamin iteration, 87
Brouncker, 13, 188, 231
Brouwer, 30
Brown, 38
Buffon, 39

C/C++ program
- π approximation using dartboard algorithm, 40
- π approximation using test for coprimality, 42
- BBP algorithm, 123, 124
- fast Fourier transform (FFT) multiplication, 141
- fft() fast Fourier transform, 143
- obfuscated (Roemer), 36
- obfuscated (Westley), 35
- sft() slow Fourier transform, 140
- spigot algorithm for π (1 digit per pass), 81
- spigot algorithm for π (4 digits per pass), 83
- spigot algorithm for π, short form, 37
- spigot algorithm for e, 85
Cadaeic cadenza, 45
Carr, 106
Carroll, 46, 72
Castellanos, 59, 60
Catalan, 229
Cataldi, 64
CECM, 113
Ceulen see Ludolph van Ceulen 182
CfToNumber(), 66
Chandah-sûtra, 121
Cheops pyramid, 211
Chephren pyramid, 211
χ^2 test, 23, 29
China, 176
Choresmia, 180
Chronicles, Book of, 169
Chudnovsky, D. and G., 1, 2, 33, 110, 111, 201, 206, 230
circle, squaring the, 7, 170, 196
Clausen, 205
Cohen, 150
compass and straight edge, 7, 98, 170

computer algebra system, 118
computers used in world records, 206
Comtet, 230
continued fraction, 64–68
- γ, 67
- ϕ, 66
- π, 32, 53, 67, 188, 231
- π^e, 67
- $\sqrt{2}$, 66
- $\sqrt{3}$, 66
- $\sqrt[3]{2}$, 67
- $\sqrt{\pi}$, 67
- e, 32, 67
- e^2, 67
Cooley, 137, 199
Cooley-Tukey algorithm, 199
coprimality and π, 41, 42
Cotes, 190
coupled Newton iteration, 147
Cox, 95
Crandall, 230
Cusanus, 183, 184

Danielson, 199
Dante Alighieri, 56, 180, 205
dartboard algorithm, 39, 40
Dase, 194, 195, 205
DATA statement, 155
Decerf, 45
definitions of π, 8
Descartes, 184, 235
Dichampt, 206
Dilcher, 157
division, 145
DNA, 4
Dodgson, see Caroll 46
Dudley, 8
Duerer, 181

$\sqrt{e}$, 67
e, 11, 14, 25, 32, 37, 60, 67, 84, 85, 116, 189, 190, 202
- series for spigot algorithm, 84
e mathematics, 116
e^2, 67
educational poems, 44–47
Egypt, 69, 167, 205, 210
elastic curve, 96, 97
Eratosthenes, 38
error distribution curve, 10
Escott, 73, 232
Euclid, 170
euclid(), 42
Euclidean algorithm, 42

Euler, 10, 11, 13, 64, 70, 74, 97, 100, 166, 186, 188–190, 194, 224–226, 230–232
– formulae, 224–226
Euler's famous π theorem, 14, 61, 190
Eulerian numbers, 157
exhaustion method, 169
expm(), 124
exponentiation, binary, 120, 121

Fagnano, 97, 98, 234
Felton, 197, 206
Ferguson, 50, 205
Fermat, 6
– Last Theorem, 7
Feynman, 3
Feynman point, 3, 46
FFT fast Fourier transform, 137
FFT multiplication, 16, 137
Fibonacci, 75, 180, 205
Fibonacci numbers, 75, 234
Filliatre, 206
Flammenkamp, 37
formula collection π, 13–14, 223–238
FORTRAN, 37, 155, 202
Fourier, 137
Fox, 39
funnel, an (in)finite, 164

$\Gamma(n)$, Euler's Gamma function, 226, 234
γ, Euler's constant, 67
Gardner, 24
Garwin, 198
Gauss, 57, 73, 87, 137, 184, 193, 195, 197, 199, 200, 232, 236
– AGM algorithm, 90, 91, 102, 114
– – derivative (Borwein), 114
– – Schönhage variant, 93
– and π, 87–102
– error distribution curve, 10
Gauss-Legendre method, 87
Gauss_AGM(), 93
Genuys, 197, 206
Girgensohn, 48
Gnaedinger, 210
Golden ratio, ϕ, 60, 66, 68, 76
Goodwin, 212
Gosper, 32, 104, 202, 206, 229
Goto, 47
Greece, 169
Gregory, 69, 70, 184, 188
– Series, 70, 189

Grienberger, 183, 205
Guilloud, 206

Haible, 215
Hardy, 107, 108, 214
Heath, 257
Hermann, 232
Hermite, 196
hfloat library, 94, 148, 247–254
high-performance algorithms, 16, 198
Hilbert, 213
Hiram of Tyre, 169
historical notes, 209–214
history of digit extraction records, 207
history of π, 165–207
history of π in the computer era, 206
history of π in the pre-computer era, 205
Hofstadter, 15
holy triangle, 211
Hutton, 73, 232
Huygens, 64, 192
hyperspheres, 158

Imhotep, 211
India, 168, 179, 185, 205
individual digits of π
– any base, 128
– hexadecimal base, 118
Internet π clubs, 11
Intuitionism, 30
IOCCC – International Obfuscated C Code Contest, 36
irrationality of π, 5, 65, 192

j Function, 27
Jacobi, 95
Japan, 194
Java program
– spigot algorithm for π, 77
Jeans, 44
Jeenel, 206
Johnson, 198
Jones, 165, 193
The Joy of π, 201
Jugend forscht, 63
Jyesttha-devan, 186

Kabus, 63, 64
Kanada, 1, 11, 12, 17, 20, 28, 114, 200, 206
Kanigel, 105
Karana-Paddhati, 185
Karatsuba, 132

Karatsuba multiplication, 134
Keith, 45
Kempermann, 234
Khintchine, 68
Khintchine constant, 68
Kings, Book of, 169
Klein, 213
Klingenstierna, 73, 232
Knopp, 228, 233
Knuth, 28, 120, 121
Koenig, 199
Kumbakonam, 106

Lagny, 45, 193, 205, 228
Lagrange, 95
Lam Lay-Yong, 176
Lambert, 5, 65, 192
Lanczos, 199
Landau, 10, 213
Lange, 33, 231
Legendre, 57, 192
Lehmann, 205
Leibniz, 70, 188, 191, 192, 228
– Series, 70, 157
lemniscate, 95–99
Leonardo da Vinci, 181
Leonardo of Pisa (Fibonacci), 75, 180
Lévy, 68
library
– hfloat, 247–254
Lievaart, 37, 85
Lindemann, 6, 197
Liu Hui, 177, 183, 205
Liu Xin, 176
Loney, 73, 106, 232
Lord, 100
Ludolph van Ceulen, 182, 205
Ludolphian number, 153, 183

m-zero, 201
Machin, 13, 72, 73, 166, 192, 205, 232
Machin formula, 72
Madhavan, 186
Maple (computer algebra system), 118
Maple program
– Leibniz Series, 157
Mathematica (computer algebra system), 35, 111, 118, 127
Matiyasevich, 234
Matsunaga, 194, 205
Metius, 181
Miyoshi, 206
mnemonic verses, 44–47

modular equations, 57, 105
modulo exponentiation, binary, 120
Monte Carlo methods, 39
Morgan, 196
multiplication, 16, 131–145, 198
– FFT, 137
– Karatsuba, 132
– school method, 131
– using Fourier transforms, 137
– using logarithms, 136
MuPAD (computer algebra system), 118
MuPAD program
– `Sqrt Coupled Newton`, 148
– `sumalt`, 151

Nemorarius, 184
new goals, 19
Newton, 104, 145, 188, 205, 228
Nicholson, 206
Nilakantha, 186, 205, 223, 224
– formulae, 223–224
Niven, 22
Noether, 213
normality of π, 4, 21–34
North, 156
nth root calculation, 149
`NumberToCf()`, 66
numerical integrals, 95
numerology, 25, 154

obfuscated C see IOCCC 36
Olds, 65
Osler, 161, 230
Oughtred, 166

Parameswaran, 186
Peirce, 14
Percival, 20, 128, 203, 207
performance index of arctan formulae, 73
Perron, 65
ϕ (Golden ratio), 60, 66
π, 1–265
– Symbol, 165
π AGM formula, 87
π Clubs, 11
π Continued Fraction, 53, 231
π Mathematics, 116
π Room in Paris, 50
π and ϕ, 60
π and e, 11, 14, 60
π calculations on the Internet, 215–221
π definitions, 8

π formula collection, 13–14, 223–238
π law, 211
π quiz, 153
π series from India, 15th century, 186
π, Digits 0 to 5,000 (base 10), 240–241
π, Digits 0 to 5,000 (base 16), 242–243
π, predictions, 198
π = 2, 155
Pi and the AGM, 115
Pi: A Source Book, 116
PI *see* performance index of arctan formulae 73
Pi Trivia Game, 153
PiHex project, 20
pispigot.htm, 77
Plato, 56, 170, 205
Plouffe, 19, 47, 48, 60, 117, 128, 207, 227, 235
Poe, 45
Poeppe, 210
polygons, 170, 175
positions, self-referential, 3
Preston, 34, 201
prime pairs, 7
Pringsheim, 65
probabilities, 21
probability subjectivism, 21
Pschill, 227
PSQL algorithm, 118
Ptolemy, 176, 205

Rabinowitz, 37, 77, 82
radius, 166
radix conversion, 79
Ramanujan, 13, 27, 57, 58, 104, 105, 199, 200, 214, 226, 227
– biography, 105–109
– formulae, 226–227
– squaring the circle, 58
Ramanujan and π, 103–111
random number generator, 27, 40
random numbers, 44
randomness of π, 21–34
Regiomontanus, 181
Reitwiesner, 206
residual sum, 186
Rhind Papyrus, 167
Rogers, 65
Romanus, 182, 205
Rome, 176
rounding up, approximation through, 38
Rummler, 162

Runge, 199
Rutherford, 194, 205

Sadratnamala, 185
Sagan, 14
Salamin, 87, 102, 200, 231, 236
Salomon, 169
Sastras, 185
Schönhage, 92, 236
series calculation, 150
Shakespeare, 46
Shanks, D., 49, 62, 197, 206
Shanks, W., 50, 195, 205
Sharp, 189, 205, 228
shortcuts to π, 35–50
sieve procedure, 38
Simon Fraser University, 113
Singmaster, 213
slide rule, 136
Smith, 205
Snell, 50, 178, 183, 205
spigot algorithm
– faster variant, 82
– pseudocode, 80
spigot algorithms, 77–85
Spring, 107
square root calculation, 146
squaring the circle, 7, 170, 181, 196
– with holes, 162
state of pi art, 1–20
Stern, 169, 231
Stieltjes, 67
Stifel, 196
Stirling, 61, 97, 229
Strassnitzky, 194, 232
Størmer, 73, 197, 232, 233
Sulvasutras, 168
sumalt algorithm, 150
symbol π, 165

tables, 239–245
Takahashi, 1, 206
Takebe, 194, 230
Tamura, 206
Tanta Sangraham, 185
tertium non datur, 30
theta functions, 95
time needed by world records, 206
transcendence of π, 5, 197
Tropfke, 169, 171, 180
Tsu Chhung-Chih, 5, 178, 205
Tukey, 137, 198

Umasvati, 169

universe, 4, 17, 153

Vega, 193, 205
Viète, 13, 160–162, 182, 187, 205
Villegas, 150

Wagon, 19, 77, 82, 126, 203, 227
Waldo, 212
Walker, 107
Wallis, 9, 64, 161, 163, 187
- product, 9
Wan Mang, 176
Wang Fan, 177, 205
waves in the π sequence, 34
Wedeniwski, 12
Wells, 190
Westley, 35
Wetherfield, 232
Whish, 185

Wiles, 7
Williams, 26
Winter, 37
world records, 12
- list of, 205–207
Wrench Jr., 197, 205, 206, 232
WWW, 11

Yoshino, 206
Yukti-Bhasa, 185
Yukti-Dipika, 185

Zagier, 150
zeta function $\zeta(s)$, 234
Zhang Heng, 56, 176, 205
Zimmermann, 8
Zoser pyramid, 211
Zu Chongzhi *see* Tsu Chhung-Chih 178

Printing: Saladruck, Berlin
Binding: H. Stürtz AG, Würzburg